Energy Management in Wireless Sensor Networks

Series Editor
Guy Pujolle

Energy Management in Wireless Sensor Networks

Youcef Touati
Arab Ali-Chérif
Boubaker Daachi

First published 2017 in Great Britain and the United States by ISTE Press Ltd and Elsevier Ltd

ISTE Press Ltd
27-37 St George's Road
London SW19 4EU
UK

www.iste.co.uk

Elsevier Ltd
The Boulevard, Langford Lane
Kidlington, Oxford, OX5 1GB
UK

www.elsevier.com

British Library Cataloguing-in-Publication Data
A CIP record for this book is available from the British Library
Library of Congress Cataloging in Publication Data
A catalog record for this book is available from the Library of Congress
ISBN 978-1-78548-219-9

Printed and bound in the UK and US

Contents

Preface

This book addresses the issue of energy management in wireless sensor network (WSN) implementation. In this context, it remains insufficient and inadequate to seek a material solution only to guarantee efficient functioning alongside an increase in the lifetime of the network. It is therefore necessary to focus on other software solutions that allow efficient information processing upon acquisition and until the final destination by taking account of sensor characteristics, i.e. weak storage capabilities, processing power and related energy constraints. Partial fulfillment of these needs entails the development of low-consumption computational tools and formal strategies using mechanisms based on information routing technologies.

In the first two chapters, we deal with latest WSN developments, before presenting the structure and composition of a sensor node, the functional architecture of a WSN and the different choices for improving energy autonomy and conservation. We then set out the taxonomy of different technologies used for energy optimization and finish by illustrating the problem to be addressed.

In the fourth chapter, we cover the issue of routing in hierarchical architectures, particularly networks with high density. In the fifth chapter, we explore the range of routing solutions developed in the relevant literature by focusing on factors improving and/or damaging the performance of networks and highlighting their adaptability.

Chapters 6 and 7 present some formal solutions developed at the LIASD[1] research laboratory at Paris 8 University. A first adaptable routing solution implements a new non-linear energy model with a child–parent communication concept, while a second solution allows problems caused by data instability and asymmetry in communications links, particularly during the recognition phase, to be avoided. The outcomes will be evaluated in the eighth chapter on the basis of a comparative study with other existing routing mechanisms.

This book is aimed at people who are not necessarily experts in wireless sensors, and can be used by engineering students, students pursuing professional or research masters, or doctoral students in the field of new communication technologies. It may also be suitable for manufacturers wishing to develop partnerships with universities on optimizing energy and computing resources. It can also act as basic guidance for developing support courses for university lecturers and researchers.

Y. TOUATI
January 2017

1 Advanced Computer Science Laboratory of Saint-Denis (Laboratoire d'Informatique Avancée de Saint-Denis).

List of Abbreviations

AOA: Angle Of Arrival

AODV: Ad hoc On Demand Distance Vector

APTEEN: Adaptive Periodic TEEN

AQFSN: Active Query Forwarding in Sensor Networks

ASCENT: Adaptive Self-Configuring Sensor Networks Topologies

B-MAC: Berkeley MAC

CADR: Constrained Anisotropic Diffusion Routing

CH: Cluster Head

CPU: Central Processing Unit

CRC: Code Cyclique Redondant (Cyclical Redundancy Check)

CSMA: Carrier Sense Multiple Access

CSMA/CA: Carrier Sense Multiple Access/Collision Avoidance

D-MAC: Dynamic MAC

DSP: Digital Signal Processor

DSR: Dynamic Source Routing

EACHS: Energy Adaptive Cluster-Head Selection

FEED: Fault tolerant, Energy Efficient, Distributed Clustering

FLAMA: FLow-Aware Medium Access

GAF: Geographical Adaptive Fidelity

GBR: Gradient-Based Routing

GDIR: Geographic Distance Routing

GEAR: Geographic and Energy Aware Routing

GMRE: Greedy Maximum Residual Energy

GOAFR: Greedy Other Adaptive Face Routing

GPS: Global Positioning System

GPSR: Greedy Perimeter Stateless Routing

GRF: Geographic Random Forwarding

H-PEGASIS: Hierarchical-PEGASIS

HEED: Hybrid Energy-Efficient Distributed Clustering

HHRP: Hybrid Hierarchical Routing Protocol

HRP-DCM: Hybrid Routing Protocol based on Dynamic Clustering Method

ISO: International Standards Organization

LEACH: Low Energy Adaptive Clustering Hierarchy

LEACH-H: Low Energy Adaptive Clustering Hierarchy-Hybrid

M-LEACH: Multi-hop LEACH

MAC: Medium Access Control

MECN: Minimum Energy Communication Network

MFR: Most Forward within Radius

MULE: Mobile Ubiquitous LAN Extensions

NiMH: Nickel-Metal Hydride

MN: Member Node

OSI: Open Systems Interconnection

PEGASIS: Power-Efficient Gathering in Sensor Information Systems

QoS: Quality of Service

RR: Rumor Routing

RSS: Received Signal Strength

RSSI: Received Signal Strength Indication

S-MAC: Sensor MAC

SAR: Sequential Assignment Routing

SGNFD: Stateless Geographic Non-Deterministic Forwarding

SMECN: Small Minimum-Energy Communication Network

SOP: Self Organizing Protocol

SPIN: Sensor Protocols for Information via Negotiation

T-MAC: Timeout MAC

TBF: Trajectory-Based Forwarding

TDMA: Time Division Multiple Access

TDOA: Time Difference Of Arrival

TEEN: Threshold-sensitive Energy Efficient sensor Network protocol

Tiny-OS: Tiny-Operating System

TL-LEACH: Two Level-LEACH

TOA: Time Of Arrival

TOSSIM: TinyOS-SIMulator

TRAMA: TRaffic-Adaptive Medium Access

UOV: Unit of Value

V-LEACH: Vice-LEACH

WBAN: Wireless Body Area Networks

Introduction

Technological advances connected to the miniaturization and integration of electronic components and to computer programming have brought about drastic changes in the field of wireless networks, giving rise to a new generation of small sensors that are able to operate independently and interact according to established communication protocols, as happened in WSN. These sensors operate around a dedicated OS and have similar functions to those of a traditional computer with microcontroller, transducer/actuator and transmitter/receiver components.

The fields of application are numerous and can include detection and environmental surveillance, transport management, traffic control and intelligent spaces, industry, health, home automation, the military, space and so on. In health-related applications, for example, the use of a WSN can improve the quality of care by using surveillance and monitoring in patients' homes. This allows medical personnel to make diagnoses quickly and therefore plan accordingly for any subsequent operations. There is also a type of advanced WSN, i.e. WBAN[1] or physical networks, widely used in the field of e-health, where data collection is carried out through the implantation of microsensors on targeted parts of the human body, as in electrocardiograms of electroencephalograms, for example.

1 Wireless Body Area Networks.

A WSN can be deployed specifically in structured environments or randomly in hostile ones which makes it vulnerable to multiple failures, ranging from physical defects provoked by environmental factors to a lack of energy resources caused by exhausted battery devices. A human intervention is generally difficult, or almost impossible, to carry out because of sensors' locations. Consequently, energy consumption management remains an unsolvable problem when designing and implementing WSN. It remains inadequate to guarantee efficient functioning alongside an increase in network lifetime by seeking only a material solution. It is therefore necessary to turn to other software solutions that allow information use to be controlled from the acquisition to its final destination by taking into account innate characteristics of sensors, i.e. weak storage capabilities, processing power and related energy constraints. Partial fulfillment of these needs entails the development of low-consumption computational tools and formal strategies applying mechanisms based on information routing technologies.

It should be noted that, in this book, a part of the proposed work has been addressed in the context of a PhD thesis [AOU 15b] that Mr. Touati and Mr. Ali-Chérif have supervised at the LIASD research laboratory at Paris 8 University.

1

Energy Management in Wireless Sensor Networks

Over the last few years, the technological advances in wireless sensor network (WSN) applications have sparked great curiosity and a growing interest among both users and manufacturers, as well as in the research community. Significant challenges have been overcome to ensure their implementation by addressing problems arising from deployment and connectivity, and from routing and securing information, although much remains to be done at the energy management stage. A WSN is made up of a set of sensor nodes, using supply devices or batteries to operate and interconnected via radio links to ensure data reception, processing and transmission. Increasing the autonomy of sensors and extending the network lifetime can therefore be considered among the main objectives by examining interesting methods and studies that optimize energy consumption, and suggesting mechanisms to improve it. These mechanisms can involve several action levels which can range from the deployment stage to the information exploitation stage.

This chapter briefly describes the latest developments in WSNs and presents the structure and composition of a sensor node, the functional architecture of a WSN and guidelines for enhancing autonomy and energy conservation.

1.1. Introduction

Wireless sensor[1] networks are a type of ad hoc network comprised of mobile and/or static sensor nodes capable of being deployed in known or unknown environments. These sensors have an energy capacity allowing them to operate independently and intelligently, and to communicate via radio link according to established routing mechanisms [YIC 08, AKY 02, CHO 03, TUB 03]. Unlike ad hoc networks concerned with guaranteeing a better quality of service, i.e. bandwidth and the transmission delay [PIR 11, ULE 06, SAN 05], WSNs promote the optimization of energy constraints as they have limited resources in term of energy, data storage and calculations. This focus is shown through the implementation of mechanisms allowing the network's lifetime to be extended (Table 1.1).

Characteristics	WSN	Ad hoc networks
Network density	High	Low/Average
Likelihood of interference	High	Low/Average
Division of network	Possible	Unlikely
Resources	Limited	Acceptable
Type of communication	Distribution	Peer to peer
Addressing	Geographic localization	IP addressing
Energy source	Irreplaceable	Replaceable
Duplication	Highly likely	Unlikely
Breakdown	High risk	Low risk

Table 1.1. *WSN versus ad hoc network*

A WSN is in fact composed of a number of nodes with the same roles, deployed in a structured or random operational environment at a high risk of breakdown. Breakdowns can stem from a lack of energy resources (battery attrition) or physical faults caused by environmental factors (rain, wind, etc.), which can complicate data transmission. Human involvement is at times absent and therefore unable to provide solutions (i.e. replace used batteries), although this is not the case with ad hoc networks where mobile nodes are built in to resolve breakdown issues. It would thus be very

1 A sensor is an extremely small device with very limited resources (energy, memory), which is autonomous and able to acquire, process and transmit information, using radio waves, to another entity (sensors, processing units, etc.) over a distance of several meters.

interesting to turn to the use and implementation of high-level strategies, such as routing, which could allow malfunctions to be resolved.

A high density roll out in WSNs encourages the use of multi-hop communication involving, on the one hand, low energy consumption and, on the other hand, consideration of issues linked to signal propagation which are often encountered in long-distance wireless transmissions. There are currently WSN-type applications that require the implementation of hundreds or even thousands of sensor nodes with the aim of ensuring both sufficient network coverage and, in particular, the ability to resolve breakdowns. The likelihood of there being one or several neighboring nodes able to take over from a non-operational node is therefore very high. In other words, if nodes randomly deployed in a given environment are distributed uniformly, the risks of network interruption or separation are lower than with an ad hoc network. Potential problems of communications interference or data duplication can also be managed.

Generally, a WSN includes sensors that interact with each other via radio links (Figure 1.1). Information that emerges due to external events or user requests can reach the desired destination along various channels. The aim is therefore to find the best route between a source and its destination (i.e. a base station) by optimizing performance criteria, such as the use or resources and/or the quality of service.

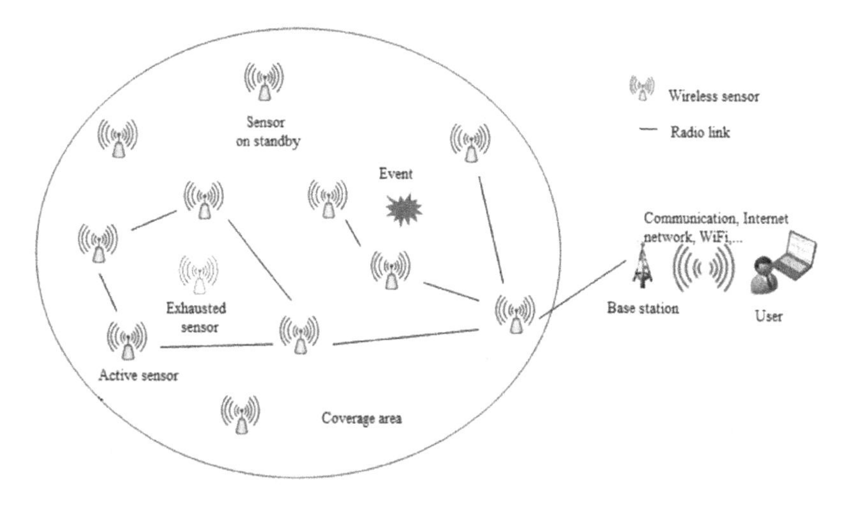

Figure 1.1. *WSN overview*

Optimal management of these resources is essential to increasing the operational lifetime of a network. The idea of an operating life has been widely studied and is subject to various interpretations linked to the functioning of the network. For instance, a network can be described as inactive due to several considerations, from the loss of one node or exhaustion of 50% of the network, to the disappearance of a detection area.

WSNs have a functional architecture, which ensures the interconnection of multiple sensors using an OSI[2] model established through the standard ISO[3] (Figure 1.2). This model allows communication between different layers to be standardized using relevant protocols [SHI 01, POT 00, KAH 99].

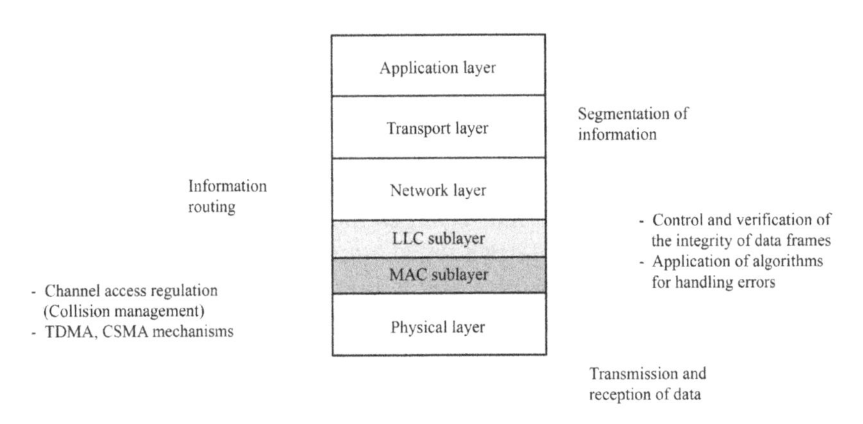

Figure 1.2. *Protocol stack of a WSN model*

Each layer has its own protocol and communicates with the other adjacent layers using its mechanism to exchange information. In this way, the transport layer structures the information emerging from the application layer in the form of segments so that they can reach the network layer. In the other direction, it reorders and then rebuilds the different segments from the network layer to direct them to the application layer. The network layer identifies the channels for the different data segments in a frame using information routing mechanisms. The LLC and MAC sub-layers form the

2 Open Systems Interconnection.
3 International Standards Organization.

link layer. The former allows the integrity of the data frames received to be controlled and verified, as well as allowing transit to the physical layer by applying error handling algorithms, i.e. CRC[4]. The MAC sub-layer is primarily concerned with regulating the access channel and standby mode, by using technologies such as TDMA[5] and CSMA[6]. The physical layer takes care of data transmission and reception.

There are numerous protocols operating at different levels of the stack with choice and implementation; however, they remain dependent on the application context (i.e. operational environment and network topology) and on the issue of making the network's operational time sustainable (i.e. managing energy consumption and optimizing resources). Implementing an energy management plan is essential to ensuring that each sensor, and the network more broadly, functions well.

Going forward, we will present and examine the idea of optimizing energy consumption, and show the taxonomy of various existing optimization techniques before setting out the issue to be addressed.

1.2. Energy consumption in WSNs

Optimizing the use of resources, particularly energy resources, in WSNs depends not only on the operational status of sensors, but also on the routing and data security techniques applied [LI 11, ALK 04, ASL 02, SHA 02, SCH 01, HEI 00]. As shown in Figure 1.3, the operational level of a sensor can correspond to several operational states: capture, processing, communication and standby.

The communication phase incorporates transmitting, receiving and listening. It is a stage in which a sensor node develops the most data routing capability. The more distant the nodes, the more energy is consumed. Conversely, during the capture and processing stages, the sensor uses less energy. It is the same with standby, where a background task continues to operate.

4 Cyclical Redundancy Check.
5 Time Division Multiple Access.
6 Carrier Sense Multiple Access.

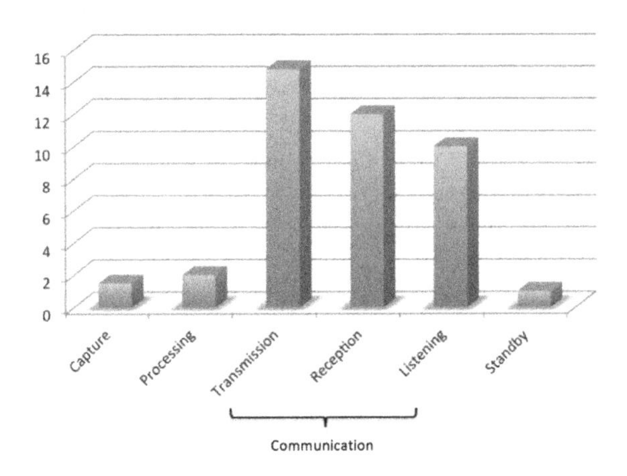

Figure 1.3. *Sensor's levels of energy consumption*

Playing just as important a role as software, the hardware architecture of a sensor node can largely provide solutions, particularly at the processing, capture and standby stages [AKY 02, PAO 07, TIL 02]. In this way, it is possible to distinguish four main units (Figure 1.4): a capture unit, which enables real data (temperature, humidity, pressure and so on) to be acquired in the form of analog signals, and then conditioned and converted into usable data. Some sensors can integrate DSP[7] signal processing and measuring cards for various uses. The energy expended during data acquisition is largely due to conditioning and converting signals.

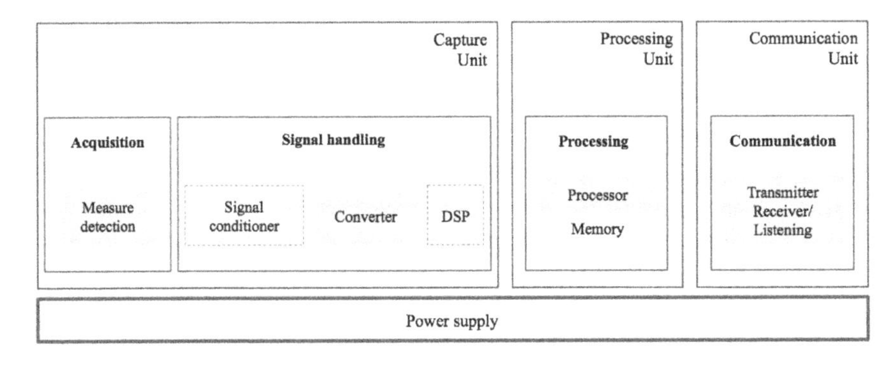

Figure 1.4. *Physical architecture of a sensor*

7 Digital Signal Processor.

At the level of the processing unit, the data acquired are exploited using a processor that performs calculations, including aggregation or synchronization with other sensors. This unit has memory for data storage and two interfaces for communicating with other units. Based on its operating system (i.e. TinyOS[8]), it allows all modules to be managed and the various tasks required to acquire and transmit data to be organized. The energy used by the processor is undoubtedly more important than the energy used during data acquisition, but less so than the energy used during communication. The communication unit, for its part, handles all radio broadcasts and reception, using relatively significant amounts of energy in comparison to the acquisition and processing units. In Pottie *et al.* [POT 00], it was shown that transmitting a single bit via the radio module requires as much energy as carrying out several thousand instructions in a program in a processing unit.

The radio module continues to use a lot of energy despite enormous efforts made in recent years in the fields of micro- and nanotechnology. The implementation of such a module requires the consideration and examination of different functional modes in the operational environment. It has four modes: active, receiver, transmitter and sleep. The active mode of a sensor node is equivalent to turning on the radio without transmitting or receiving data thereby entailing an additional useless loss of energy. In order to avoid this, the sensor node can sleep by disconnecting its radio, and the transition from sleep mode to active mode requires restarting several software programs and components, using more energy than if the sensor node remained in active mode. Energy management can therefore be handled using a protocol from the MAC sublayer. In the sender/transmitter modes, the sensor transmits or receives messages in the form of bytes. The bigger the message, the larger the number of bytes and, consequently, the greater the amount of energy used. There are two proposed approaches for resolving this [ANA 09, ANS 08, KES 06, GU05, MIR 05]: the first allows the systematic transmission of information to be avoided by passing through a safeguard. Transmission is only activated in cases of need or demand, which reduces energy consumption, but uses memory. The second approach works according to the data aggregation principle by processing on the basis of mathematical functions (total, average, maximum/minimum, filtering, etc.) or algorithms to optimize the number of bytes to be transmitted. In this

8 Open-source operating system for WSNs.

solution, energy consumption can be optimized without requiring significant processing power (Figure 1.5).

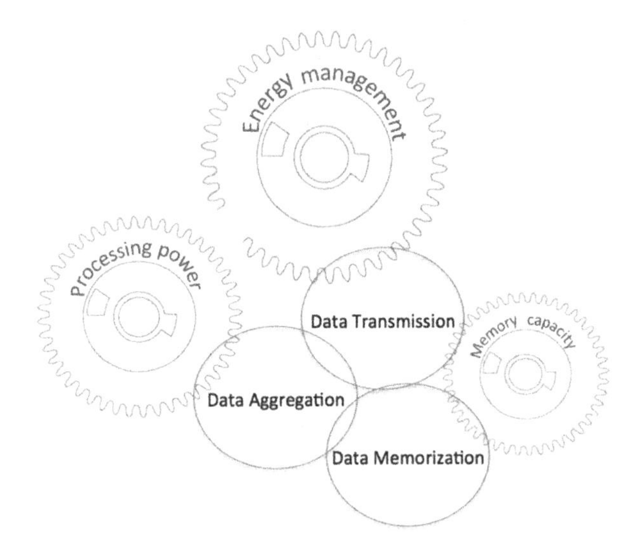

Figure 1.5. *Distribution of energy consumption*

Wireless sensor autonomy is managed for the most part by the power module or battery, which provides the necessary energy for operation (Figure 1.4). Choosing the type of power supply depends on the physical characteristics of the sensor used and the amount of processing to be carried out. It is therefore essential to estimate the amount of energy necessary in order to set up some kind of application, as the lifetime of a sensor can vary between several months or even a year without outside interference. Turning towards renewable and natural energy sources, such as solar and wind power, is proving to be highly useful. However, in many situations, the deployment environment can be ill-suited to using these sources due to existing restrictions. For instance, the use of miniaturized solar panels is very interesting but can be less effective in Saharan desert environments due to climatic changes (sand storms), which can directly and negatively influence and affect their operation. The most economical solution currently recommended is a power supply of standard lithium, alkaline or NiMH[9] batteries.

9 Nickel-Metal Hydride.

2

Optimization Techniques for Energy Consumption in WSNs

In recent years, several pieces of research have been undertaken to study and address energy consumption in WSNs. All of these works are based on different parameters relating to the operational mode of the sensor, mobility, quality of service, routing and securing information, and so on. Therefore, on the basis that radio transmission requires greater amounts of energy to route information to a destination, undertaking such research is useful for designing and developing new mechanisms that will provide solutions to improve energy efficiency. Depending on the context, these solutions can be categorized according to three different techniques [ANA 09] (Figure 2.1): the partition of operating time, the structure of data and the mobility of sensors.

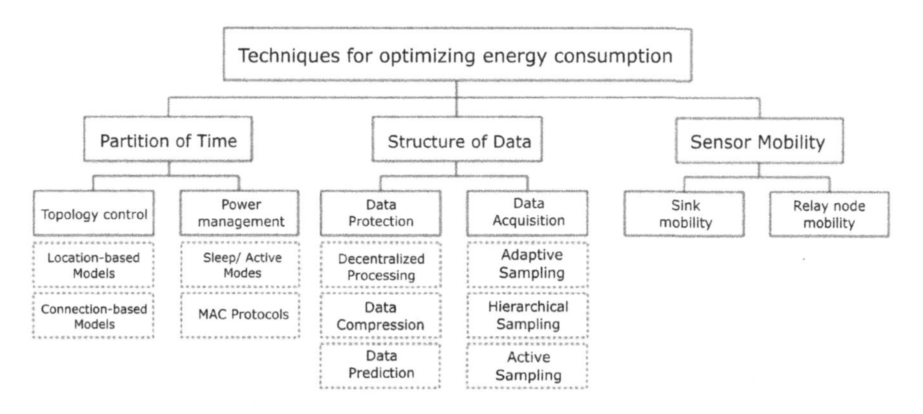

Figure 2.1. *Taxonomy of energy consumption techniques*

The choice of one technique over another is dependent on the type of network application to be carried out, the spatial and/or temporal restrictions to be satisfied and the overall operational context.

2.1. Management and partitioning of time

Commonly known as duty-cycle techniques, these techniques consist of optimizing energy use by switching the states of sensor nodes to sleep mode when no communication is required. Ideally, the radio system should be turned off when data is not being transmitted or received, and should be activated when new data packets are available, with nodes alternating between active and sleep periods in accordance with network activity. Time partition can be determined by inspecting the topology of the network or by managing the power use (Figure 2.2). It is possible to exploit the duplication of nodes by adaptively selecting minimum subsets of active nodes to guarantee connectivity. Other nodes can remain in sleep mode. Finding the optimal subset of nodes able to ensure connectivity falls within topological control, with the idea being to exploit duplication in the network in order to increase its lifetime by a factor of 2 to 3 compared to a network using all of its nodes [GAN 04, MAI 02, WAR 07].

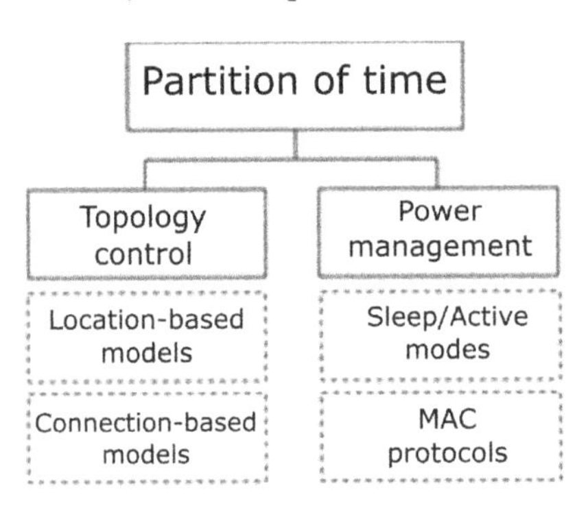

Figure 2.2. *Taxonomy of techniques based on time partition*

Within the category of topology control protocols, location-based models determine the sensor nodes that can be activated, as well as the activation time, in accordance with their positions. It is assumed that these nodes will already be known from existing uses, for example in GPS[1]. The GAF[2] protocol is part of this and allows energy consumption to be optimized while maintaining an acceptable level of routing quality (packet loss, message latency) [XU 01]. The detection area is divided into small adjacent and equivalent virtual zones in which a single node can be activated at a precise moment. This moves through the coordination and distribution of messages to all nodes in each zone in order to determine the specific period of time in which they need to be active or asleep (Figure 2.3). The leader node periodically rebroadcasts a discovery message allowing other nodes to be activated. The network can therefore be underused overall, as the protocol is based on communication between adjacent tables using half of the radio range.

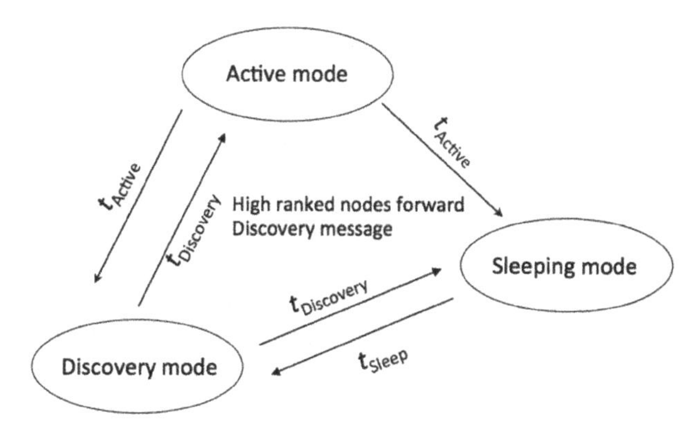

Figure 2.3. *State transitions in the GAF protocol*

In the discovery mode, a node exchanges discovery messages including grid identifiers following other nodes within the same grid. It can therefore be a master if it does not hear any other discovery message for a predetermined duration $t_{\text{Discovery}}$. If more than one node needs to be a master, it can be the node with the longest expected lifetime. In this way, the master remains active in order to handle routing for t_{Active}. After that, the node enters discovery mode to give other nodes within the same grid an opportunity to

1 Global Positioning System. As GPS is very expensive to embed in every sensor node, it can only affect a limited number. This is enough to determine the positions of all sensor nodes.
2 Geographical Adaptive Fidelity.

become a master. In scenarios with high mobility, sleeping nodes should wake up to take over the role of a master node, where the sleeping time t_{Sleep} is computed according to the estimated time the nodes remains within the grid. These transitions depend on the type of application and the related parameters to be tuned during the routing process. Similarly, the GeRaF[3] protocol also uses the position of nodes [CAS 05, ZOR 03a, ZOR 03b] and enables sensor nodes to be in listening mode before changing periodically according to operating cycles, alternating between active and sleep modes. When a sensor node wishes to transfer data, it enters active mode and transmits a packet containing its own location and that of the receiver. The latter responds with an acknowledgement message in order to begin transmission. The zone around the receiver is divided into areas with different priorities. The closer the area is to the destination, the higher its priority. If all nodes in an area are in sleep mode, an attempt to cross another high priority area is made, and so on.

Conversely, connectivity-based models dynamically activate and deactivate sensor nodes by guaranteeing both connectivity and network coverage [KON 07]. There is the SPAN protocol, which enables a dynamic choice of coordinator nodes to be made using information provided by neighbors and connectivity [CHE 02]. These nodes, in active mode, actively participate in routing, while the majority of other nodes are in standby mode. In order to ensure a sufficient quantity of coordinators, an eligibility rule is used. In the ASCENT[4] protocol [CER 04], a node can choose to participate in the routing process or to sleep, on the basis of information on density (i.e. connectivity) and the loss of packets following collisions (Figure 2.4).

Initially, the network only considers a selected number of active nodes while the majority remain passive. The number of active nodes gradually increases, while the number of messages lost between their source and their destination (i.e. sinks) decreases. If the number of active nodes decreases, the destination solicits neighboring nodes. These transition from passive to active mode and report their availability with an announcement message. This process continues until the loss ratio of messages passes below a given threshold, before resetting itself following new events on the network (a node failure) or changes to the environmental context.

3 Geographic Random Forwarding.

4 Adaptive Self-Configuring sEnsor Networks Topologies.

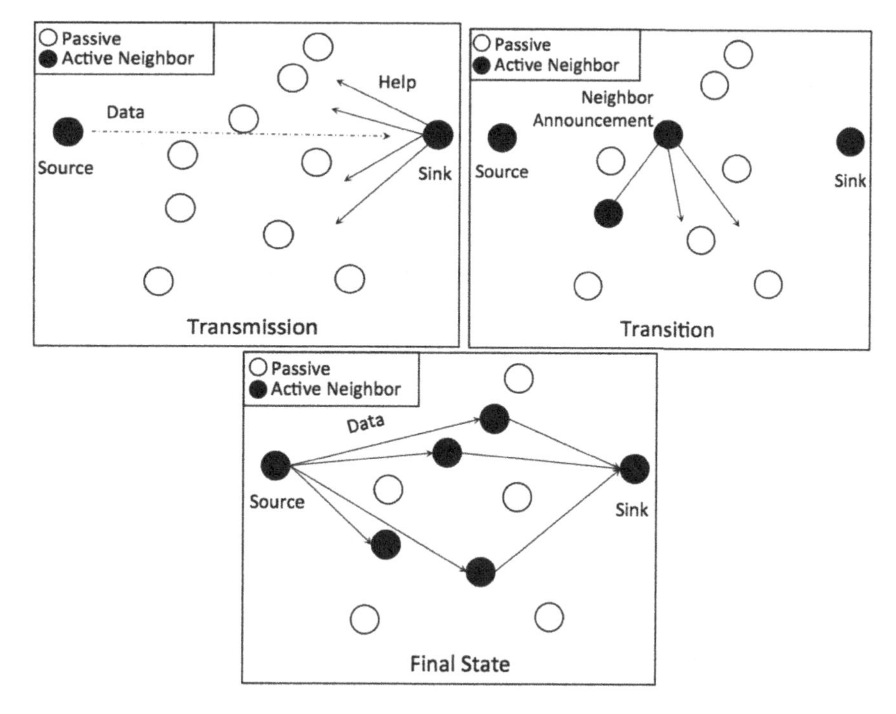

Figure 2.4. *State transitions in GAF protocol*

Techniques based on power management are divided into two large categories according to the network architecture and the level of the protocol layer. MAC[5] protocols can be identified by their reduced operational cycles and protocols based on sleep/active modes. The latter, operating beyond the MAC layer (i.e. network or application layer), provide significant flexibility as they can be adapted to application requirements and collaborate with other MAC layer protocols. These are independent protocols relating to topological aspects and network connectivity. They can, for their part, be subdivided into three main categories of protocol [ZHE 03]: on demand, scheduled and asynchronous.

5 Medium Access Control.

On demand protocols are highly effective because they allow craftily exploiting energy use, as nodes are only active during the period of time allotted to a communication phase. While the waiting time (i.e. latency) to activate a target node is very acceptable, the activation process is, on the other hand, somewhat complicated. One solution is to use a scheduled approach that allows appointments between nodes wishing to communicate to be programmed regularly and in advance. Nodes remain active for a period of time before switching to sleep mode until the next appointment. These protocols are very practical for data aggregation but require additional methods for synchronization, a process that is difficult to carry out and expensive to implement. Asynchronous protocols do not require node synchronization. Their structures are generally very simple to implement and can guarantee connectivity even in highly dynamic networks. However, active mode is regularly used, requiring several operational cycles.

The protocols operating at the level of the MAC layer optimize channel access functions by using very precise active/sleep cycles [DEM 06, KUR 07, YAD 09, YE 04]. There are three methods: the TDMA[6] method, contention methods and hybrid methods. TDMA methods [LI 04] allow cyclical access to the channel. The nodes only activate their radios during specific time slots, thereby reducing the energy used to a strict minimum. Each cluster is formed with a coordinating master node, as in LEACH[7] [HEI 00], MAC-TDMA [ARI 02], Bluetooth [HAA 00], TRAMA[8] and FLAMA[9] [RAJ 03, RAJ 05], for instance.

The contention methods allow rapid access to the channel through sleep/active methods. An example is the B-MAC[10] protocol [POL04], which is supplied with the TinyOS operating system [HIL 00] and provides basic features with a channel access mechanism, allowing an efficient optimization of energy consumption. Other more optimized versions such as S-MAC[11] [YE 04], T-MAC[12] [DAM 03] and D-MAC[13] [LU 04] have been

6 Time Division Multiple Access.
7 Low Energy Adaptive Clustering Hierarchy.
8 Traffic-Adaptive Medium Access.
9 Flow-Aware Medium Access.
10 Berkeley MAC.
11 Sensor MAC.
12 Timeout MAC.
13 Dynamic MAC.

suggested to further improve the coordination of sleep/active cycles and data collection, and in which nodes are structured in a parent–child tree. The operational time of nodes is determined in accordance with their position in the tree. Each node has a sufficient period of time to transmit data. It can also ask its parent to obtain additional slots, thereby dynamically adapting active periods in accordance with the level of traffic. With much lower levels of contention, hybrid approaches use the probabilistic aspect of the TDMA mechanism [EPH 82] to access the channel. Depending on the existing contention, the protocol can switch between CSMA and TDMA modes, hence the term "hybrid". In order to be independent of any phenomenon of a change in topology or to errors in synchronization and interference, which are widespread in dense networks, these approaches use a single-hop communication.

2.2. Data-oriented techniques

All of the techniques mentioned previously are insensitive to sampled data during capture, although such data can greatly impact energy consumption. The use of data-oriented approaches has proved attractive, as they are able to provide additional solutions to increase the effectiveness of energy optimization. Given that sampled information generally has a strong spatial and/or temporal correlation in the network [VUR 04], it is not necessary to communicate redundant information to base stations in the case of a specific cluster, for instance, which results in energy savings. Moreover, these techniques allow the amount of sampled data to be reduced, while maintaining the precision of detection subsystems, including when they have no energy.

Data-oriented approaches (Figure 2.5) can be classified according to the issue being addressed. The use of data reduction techniques allows superfluous samples to be processed, while data acquisition approaches aim primarily to reduce the amount of energy expended by detection subsystems. However, some of them can also reduce energy loss in general, which is the main objective.

Intended for specific applications, decentralized processing [FAS 07] consists of aggregating data by calculating, for example, the average of certain values at the intermediary node level in order to reduce the dimension to be transmitted. Data compression can be applied to optimize the amount of information emitted by source nodes. This technique allows the respective encoding and decoding of information at the level of data-generating nodes and the sink nodes. In the relevant literature, there are various methods that use compression [PRA 03, TAN 04, WU 03, XIO 04], but few of them relate to WSN applications.

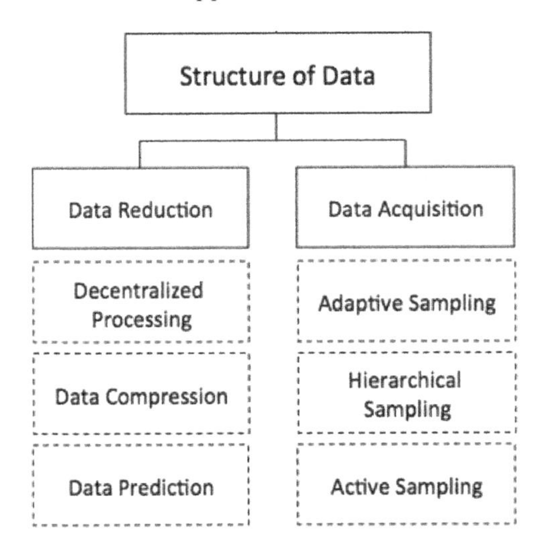

Figure 2.5. *Taxonomy of data-oriented techniques*

Data prediction consists of building an abstract model of the phenomena detected in order to describe and predict, within a reasonable margin of error, the evolution of data over time. If the measurements are sufficiently precise, then the requests sent by users can be evaluated at the sink node level using a prediction model, without requiring the acquisition of precise data from other nodes. Conversely, if the model is imprecise, it is necessary to have an explicit communication between node sensors and the sink node; i.e. real samples, such as they are, must be recovered and/or the model must be updated. In this way, the amount of information transmitted by the source nodes is reduced.

Three main types can be identified. Stochastic approaches which characterize phenomena in a random manner, in terms of probability and stochastic properties [CHU 06, KAN 08]. They are generalist and allow functions such as aggregation to be carried out at a high level but with high and heavy calculation costs, particularly through the use of simple and classical detection devices. Nonetheless, they seem to be interesting detection devices with significant calculation capabilities. It should be noted that improvements could always be made in this regard by implementing simple distribution models, and ensuring a compromise between calculations and precision. In contrast to stochastic approaches, the time series prediction explicitly considers the structure of data in order to predict future values [LI 07, TUL 06b]. This can provide satisfactory precision, even if simple ARMA models, for instance, are used. It is therefore simple and effortless to implement this on capture devices. The use of most advanced techniques [TUL 06a] does not require the contribution of all data detected as long as a model is unavailable. These techniques are able to detect and to take into account all the aberrant values as well as the inconsistencies related to the model. An interesting direction is to adopt a multi-model approach as described in [LE 07]. The third classification concerns algorithmic approaches, which must be considered case by case, and relies on heuristics or a state-transition model describing the detected phenomenon [LE 07, GOD 04, GOE 06]. They use methods or procedures to build and update the model on the basis of the chosen characterization.

The use of detection models or subsystems to collect information can, in some cases, require an overconsumption of energy in comparison to the radio, or even a higher energy consumption than that of the node sensor itself. Some sensors inherently require significant resources in order to carry out sampling, such as CCD or CMOS image sensors or even multimedia sensors. Other sensors can collect data on the phenomena detected by using active devices (i.e. sonar, radar or laser) and then transmit test signals in order to obtain information on the amount of information observed. The acquisition time can be around several hundred milliseconds or even several seconds, implying an increase in the consumption of the detection model even if the energy consumption of the sensor itself is reduced. Moreover, having high speed and high-resolution A/D converters can cause overconsumption. It is therefore necessary to modify programs by reducing the number of acquisitions, i.e. the number of samples, thereby limiting communications. Knowing that the measured samples can be correlated, techniques based on adaptive sampling are able to exploit those similarities in order to reduce the amount of data able to be acquired by transducers. For instance, if the useful data evolves slowly and without significant variations, it is then unnecessary to carry out several acquisitions. An identical approach can be applied when the phenomenon under study has not made significant developments in the zones covered by nearby nodes. Energy expended on sampling can be reduced by taking advantage of the spatial correlations of the data collected. With regard to hierarchical sampling, it appears that each sensor node has its own resolution and energy consumption. The clusters that form the network are selected and then activated dynamically in order to ensure a compromise between precision and consumption. Adopting the same strategy as the predictive approach, active sampling builds a model of the phenomenon detected using the data acquired with the aim of reducing the number of samples and thereby the consumption of energy [DES 04].

2.3. Sensor mobility-based techniques

The presence of mobile nodes in the network (including sink nodes) can improve the traffic fluidity as they allow nodes to explore the entire network, thereby evenly dividing energy consumption [LI 07]. Consequently, contention and congestion are decreased, reducing energy use. It is useful to underline the importance of the control strategies of mobile nodes to ensure better fluidity of movement [ERR 08, JUN 05a]. Indeed, problems related to the interaction of mobile nodes with the operational environment can cause physical deterioration and a failure to comply with optimal trajectories, leading to a significant loss of energy.

Two categories of mobile nodes can be distinguished; those that are part of the network infrastructure and where mobility is completely controlled using sensor nodes as robotic units, and those evolving in a deployment environment in which no controls are applied. In this way, depending on the type of mobile entity, there is an approach based on the mobility of sink nodes and an approach based on the mobility of relay nodes (Figure 2.6).

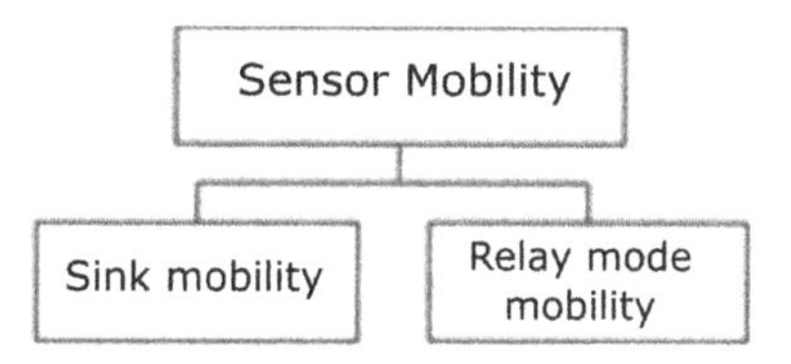

Figure 2.6. *Taxonomy of techniques based on node mobility*

All of the approaches proposed in the relevant literature that use techniques based on the mobility of sink nodes are themselves based on linear programming methods intended to optimize parameters, such as the lifetime of the network, the QoS and so on. For instance, some authors [WAN 05] propose a model that considers a mobile sink evolving in a squared grid inside a detection zone. Once contact is established, the sink node remains static for a period of time before communicating directly with its immediate neighbors. Remote sensor nodes send their messages through multi-hop transmission, with the aim of determining the optimal waiting time of a sink node. A similar model proposed in [GAN 03] using several mobile sinks has shown that the lifetime of the network can be extended by a factor between 5 and 10 in comparison to an approach implementing static sinks. An extension of the model developed in [WAN 05] that does not position sensor nodes in the detection area has been proposed in [PAP 06]. This approach considers the residual energy of sensors by adopting an appropriate routing policy to increase the performance of [WAN 05]. An approach based on implementing a distributed protocol has been proposed in [BAS 07]. It suggests a GMRE[14] structure in which a sink node selects a detection area with significant residual energy using a sentinel node chosen in each detection area. The node determines the energy using the surrounding nodes and responds to the request of the mobile sink. The latter uses this information to decide whether it must move towards the detection area.

Models that use mobile relays to collect data have been widely studied in opportunistic networks [ERR 08]. These approaches, or at least the best known, use message ferrying structures [JUN 05b, ZHA 04] with specific mobile nodes, i.e. message ferries, which are introduced into the network in order to ensure a relaying service of messages. The ferries progress through the network by collecting data using source nodes, which are transported and then directed towards destination nodes (Figure 2.7).

14 Greedy Maximum Residual Energy.

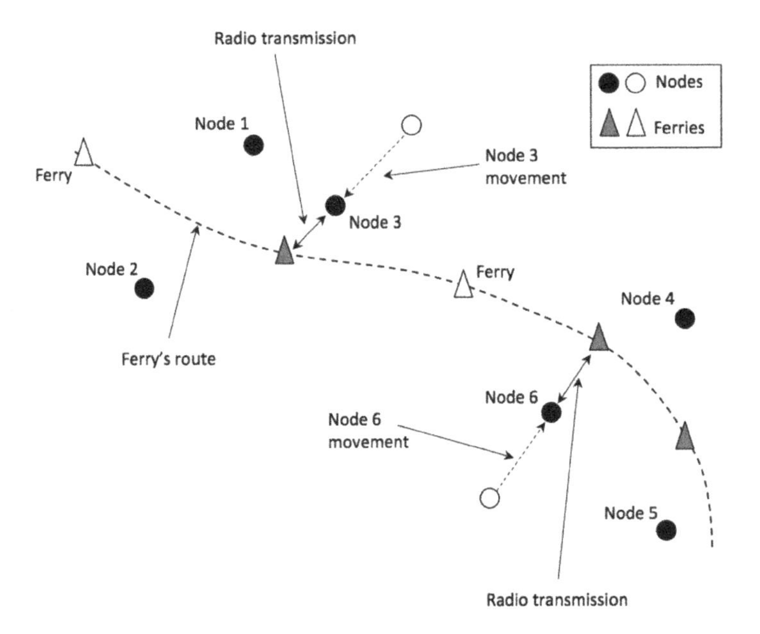

Figure 2.7. *Message ferrying in a WSN. For a color version of this figure, see www.iste.co.uk/touati/energymanagement.zip*

A similar structure (Figure 2.8) based on a three-tier operational architecture has been proposed for complex WSNs: data-MULE[15] [JAI 06, SHA 03]. The lower level consists of all nodes that periodically conduct data sampling. The intermediate level concerns the MULE mobile agent, which is able to move throughout the detection area in order to collect data being temporarily stored in the internal buffers.

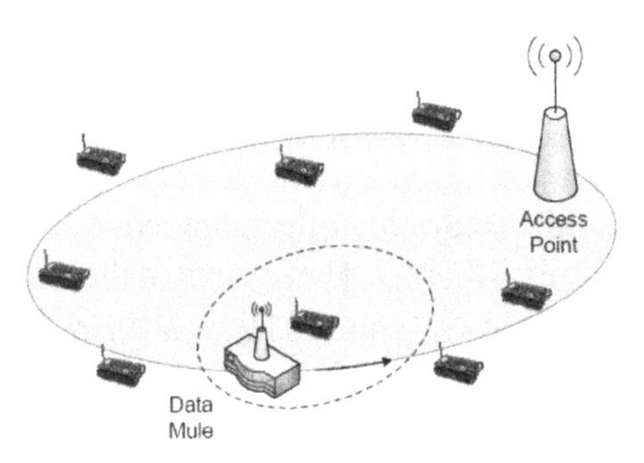

Figure 2.8. *WSN with mobile relay*

15 Mobile Ubiquitous LAN Extensions.

The MULE agent and the communications sensor use short wave radio signals, which reduce *de facto* energy consumption. In [KAN 04] and [SOM 06], mobile relays are supposed to move along the route set out in advance as it is very simple to alter the trajectory of a relay node in a WSN for various reasons, namely, the presence of obstacles and rugged terrain where autonomous vehicles can generally only move in certain directions.

Finding themselves far apart, the sensor nodes use different multi-hop routes via nodes known as hidden nodes. This is less costly in terms of energy consumption as communications are ensured via very short radio links, which is not the case in dense WSNs. Sensor nodes automatically organize themselves into clusters in which the group leaders are built to communicate through mobile relays. They are generally very close to the trajectories that mobile relays must take. All of the nodes belonging to each group transmit their data using a directed diffusion protocol to their respective group leaders, which store and then transmit the data to mobile relays.

The latency time taken by the data to reach the destination (sink node) can be considerable as the time intervals that occur between the sampling stage and the collection of data by the MULE agent, before it is routed to an access point, can be significant. Moreover, energy is lost, as all nodes must remain in active mode until the MULE agent enters their vicinity to transmit data. The idea to be considered is as proposed in [KAN 04, SOM 06, JEA 05], with several MULE agents able to cover the entire network in order to avoid substantial waiting times. One of the typical applications of this model is to develop an independent submarine vehicle to exploit and simulate the behavior of marine ecosystems [AKY 05, VAS 05]. There are other systems using the same principle of messenger agents, such as ZEBRANET, a system with wireless sensors embedded in mobile devices [JUA 02] intended to track zebras in the wild, or SWIM which is used to observe whales in the ocean [SMA 03, HAA 06].

2.4. Analysis and conclusion

We have exhaustively presented the energy conservation approaches with particular attention to solutions proposed in the relevant literature. All of these works consider not only the operational status of the sensor, but also the whole range of parameters and approaches that can influence the overall functioning of the network (data, mobility, routing, etc.) By working on the basis that the transmission stage requires greater amounts of energy to route information, research must be undertaken to conceive new techniques that allow better solutions to the challenges of energy optimization to be developed. In this regard, we have identified three main approaches, namely, the management and partition of time into operational cycles, data-oriented approaches and approaches based on sensor mobility.

Approaches based around time partition are generally unaffected by data sampled during detection and can greatly impact energy consumption. The recourse to using data-oriented approaches consequently proves more interesting as such approaches may provide additional solutions to increase the efficiency of energy optimization. These approaches process data during the information acquisition stage, thereby avoiding dealing with unnecessary samples. The amount of data sampled is reduced while maintaining the precision of detection subsystems, especially when they have run out of energy. This can be interpreted as an energy gain. Approaches based on the mobility of sensor nodes allow contention phenomena to be circumvented in order to avoid an overconsumption of energy, and therefore the premature exhaustion of related nodes. Conversely, there are significant waiting times as messages are ferried. In our humble opinion, these approaches must not be considered alternatives or competitors, but must rather be used together to take advantage of each of them.

The majority of the solutions proposed in the literature assume that the radio transmission stage uses more energy than the sampling and data processing stages. However, there are real applications where consumption is almost identical at all stages, as the sampling stage may require more time than the data transmission stage. In this way, the energy consumption of the sensor itself can also be very high. We believe that this aspect must be fully explored.

In the next chapter, we will address the issue of energy management in WSNs from the perspective of information routing. We will present the challenges and issues that allow us to structure and classify routing protocols according to the context of application and the desired objectives.

3

Routing Information for Energy Management in WSNs

The main objective in a WSN application is to guarantee the transmission of information between different sensor nodes according to a pre-established routing protocol. Improving performance in terms of longevity, connectivity and robustness requires some consideration of constraints such as energy consumption, bandwidth and the optimal use of calculation and memory resources. The conception and implementation of a routing protocol can be influenced by several factors that must be addressed before it is possible to ensure efficient data communication. Among these factors, we can refer to the deployment of nodes in operational environments. Being dependent on the kind of applications under consideration, the deployment phase can greatly affect routing performances. It can be determinist and, in this case, sensor nodes are placed manually and the data are transmitted according to pre-determined routes from source nodes to the destination. This can also be done randomly by haphazardly distributing the nodes, as in ad hoc networks. In this case, an optimal classification of the network in clusters is highly necessary in order to guarantee connectivity and enable efficient energy management. Various studies have shown that the communication between nodes in restricted areas, i.e. clusters, provides better results in terms of energy consumption and bandwidth, and information routing can be guaranteed in this case through multiple hops. Moreover, the lifetime of a sensor node is strongly linked to the lifetime of its battery [SCH 01] and the malfunctioning of one or several sensor nodes, due to battery failures or a lack of energy resources, can cause significant topological changes that require the network to be completely reorganized in order to ensure the rerouting of data. It is therefore important to consider the issue of how to manage energy consumption without a loss of performance and network precision, particularly in the case of a multi-hop WSN where each node can act as either a data transmitter or a simple router.

3.1. Challenges and issues in WSNs

In a WSN, a routing protocol must comply with the choice of information routing model from the detection stage to transmission [HEI 00]. As this is highly dependent on the type of application considered, several models have been developed and their implementation must take critical parameters into account. Among these parameters are the critical delays necessary to detect and then transmit data. These models are very well adapted to cases of applications requiring periodic data management, where nodes regularly switch from operating simply as a sensor to being a data-transmitting node, and vice versa. Other models consider transmission on the basis of establishing requests, for the base station or after events for example. In the latter case, sensor nodes react immediately following abrupt changes in the values detected. It should be noted that it is possible to combine these models.

There are applications where the networks are homogenous, i.e. sensor nodes have the same calculation and memory capacities, identical communication parameters and the same amount of energy, and are designed to carry out the same functions. However, there can be other applications where nodes have different capabilities and roles, forming a completely mixed network. This heterogeneity can raise various technical questions related to the transportation and routing of data. For instance, some applications could require a range of sensors to control the temperature, pressure and humidity, detect movements through acoustic signatures and capture images or track moving objects using video. The reading and transportation of data can be carried out at different frequencies and on the basis of several models with a minimal quality of service, as in the case of hierarchical network structures where Cluster-Head (CH) sensor nodes play a different role to ordinary nodes. Selected just after the deployment of the network and in contrast to other sensor nodes, CH nodes are more powerful in terms of energy, bandwidth and memory. Aggregation of data emerging from several different sources is an important function carried out by CHs.

It allows superfluous information to be deleted (suppression of duplicated data, minimum, maximum and average) and therefore the number of messages conveyed to the base station to be optimized. In this way, these techniques for data aggregation and fusion allow energy consumption to be considerably reduced at the level of each sensor node and across the network as a whole.

In static sensor networks, some nodes can break down due to a physical failure or very simply because of a lack of power. These breakdowns must not impact the functioning of the network under any circumstances. The data packets can follow various multi-hop paths in order to reach a desired destination (i.e. a base station, sinks). However, there can be more solicited paths or nodes than others, causing contention or congestion that translates directly into significant energy consumption and therefore the premature exhaustion of nodes [LI 07]. The presence of mobile nodes in the network (including sink nodes) can improve the traffic flow, which allows the nodes to explore the entire network and thereby distribute energy consumption evenly. By acting as data collection points every time that they are close to a source (i.e. static nodes), mobile nodes limit both the number of hops and the distance over which the information is transported. Consequently, contention and congestion are reduced and less energy is used. The models developed in this context have largely been used in the frameworks of opportunistic networks [ERR 08, BAS 07, JUN 05a, ZHA 04].

In numerous applications, efficient routing occurs primarily by optimizing the energy deployed by all sensor nodes, to the detriment of improvements to the quality of service. This is carried out in order to increase the lifetime of the network and improve the quality of data transmitted, or guarantee that it is relegated to the background. However, in certain applications, the information must be conveyed in very short periods of time in order to preserve the freshness and quality of the data transmitted. It is therefore essential to limit the latency time when issuing data in order to guarantee an acceptable quality of service. All of these challenges and concerns lead us to structure and classify routing protocols according to the desired objectives. We will go on to present the taxonomy of information routing mechanisms in the WSNs in question from the perspective of energy optimization.

3.2. Taxonomy of routing mechanisms in WSNs

The classification of routing mechanisms or protocols in WSNs can be examined from several different angles [ALK 04]. From a structural point of view (Figure 3.1), i.e. according to the topology of the network and the paradigms chosen to optimize cost functions, we can refer to data-based structures, i.e. flats where all of the sensor nodes have the same functions and play the same role when processing information. In contrast, in hierarchical structures, nodes do not have an equivalent role. According to their hierarchy in the network, they can have privileges such as performing calculations and communicating with the base station. In location-based routing, the positions of sensor nodes are used to route and convey information. Each sensor has a locating system such as GPS that allows its position to be calculated and thereby transmitted to the target node.

From a functional point of view, this depends on the type of applications implemented. There are negotiation-based routing technologies for dialog between sensor nodes, the establishment of requests, consistency, the improvement of QoS[1] through data exchange and multiple paths to enhance reliability and performance. The latter technique is very similar to techniques based on flat topologies.

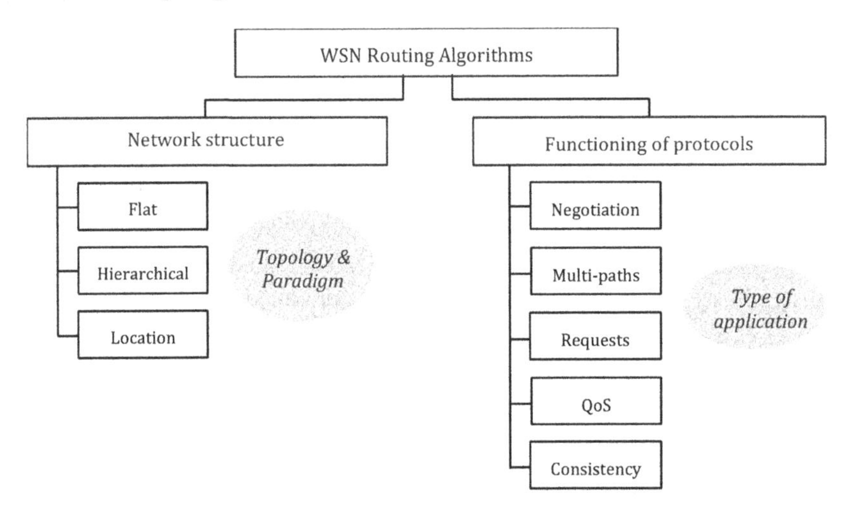

Figure 3.1. *Routing mechanisms in WSNs*

1 Quality of Service.

These routing mechanisms can also be classified according to criteria for establishing paths between source and target sensor nodes. It is possible to distinguish a category of proactive protocols with several paths without actually feeling the need to use them. In contrast to the proactive protocol category, in the reactive protocol category, the potential paths to the target are determined upon request. Hybrid protocols combine both proactive and reactive protocols. However, as the nodes are static, it is preferable to use routing tables instead of information provided by reactive protocols. In this way, the energy consumed is minimized during the path research stage and the configuration of reactive protocols can be avoided. In the case of cooperative protocols, sensor nodes convey information to a central node in order to be aggregated and then processed. This reduces the number of paths to reach the base station and therefore optimizes energy consumption.

3.2.1. *Routing based on the structure and the topology of the network*

As highlighted above, several information routing mechanisms can be distinguished according to their structural organization and the paradigms used to optimize network performance. We will now go on to examine all of the protocols implemented in this context.

3.2.1.1. *Data-centered flat routing*

In the category of multi-hop flat routing protocols, sensor nodes work together to carry out their tasks, with only the base station responsible for collecting information [SAD 03, INT 00]. As a WSN generally has tens, or even hundreds or thousands of sensor nodes, it is impossible to assign an identifier to each of those nodes. The result is therefore routing centered mainly on data, where the base station sends requests to pre-selected areas and puts itself in standby mode (Figure 3.2).

The topology related to this kind of protocol is characterized by a high fault tolerance through the construction of new paths, with low maintenance costs. Nonetheless, a significant number of control messages are used to ensure that the network functions correctly without actually having to improve scalability. Structure or topology changes, i.e. node loss due to breakdowns or energy loss, can damage network performance in some cases.

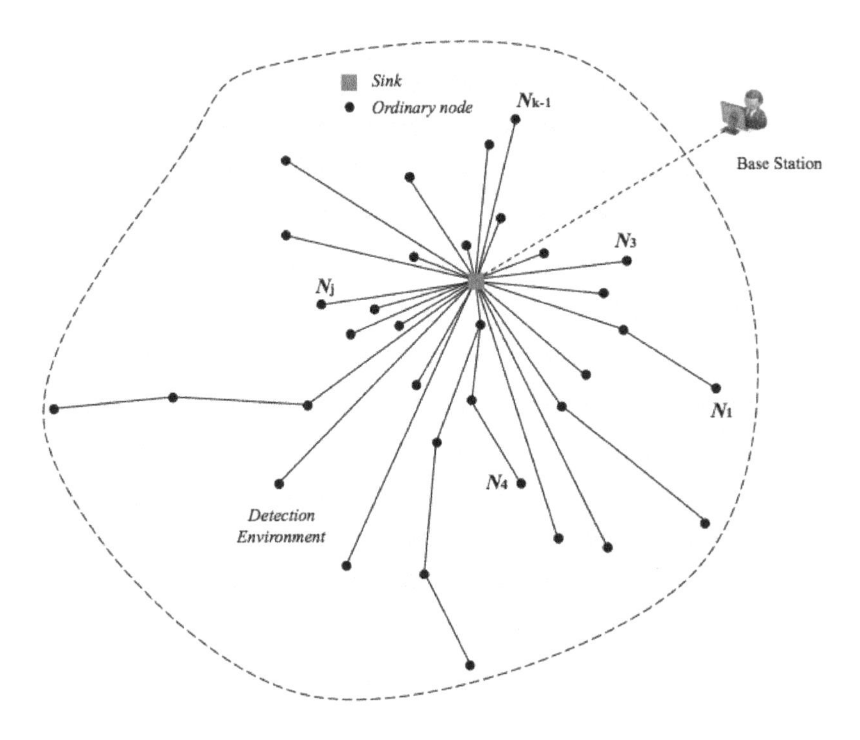

Figure 3.2. *Centralized architecture with a flat topology. For a color version of this figure, see www.iste.co.uk/touati/energymanagement.zip*

Pioneering works such as SPIN[2] [YE 01] have shown their efficiency using negotiation-based mechanisms [HEI 99, KUL 02]. The problems encountered during information transmission through flooding [JOH 96], such as duplication and redundancy of data, are taken into consideration, optimizing energy resources at the level of each sensor node.

2 Sensor Protocols for Information via Negotiation.

A simple, reactive and efficient protocol developed for ad hoc and MANET networks known as DSR[3] uses dynamic routing to determine the sequence of nodes necessary to convey data to their destination, i.e. sinks [JOH 96, JOH 01]. It manages and configures itself independently through two mechanisms: automatic discovery of paths, i.e. Route Discovery, and online path maintenance and correction, i.e. Route Maintenance. Each node has a routing table containing all of the paths to be taken. The path search works by flooding the network using Route Request message packets, with nodes receiving these packets consulting their routing tables first of all, and if the result is productive, they refer in turn to the source node and use unicast routing to send Route Reply messages containing a list of suitable nodes to follow in order to reach the destination. AODV[4], part of a reactive protocol family, uses the distance vector concept based on the stored paths in the routing tables and the sequence numbers contained in control messages [PER 99]. Path search includes the diffusion of a route search message throughout the network, which is relayed by all nodes (on a single hop) to the destination. Once the destination is reached, the latter returns a response message to the source along the same path, and records a trace of the request before distributing it in turn to all of the other nodes. In the case of a link breakdown, an error message is sent to the source and the path is destroyed and thereby removed from the routing tables of all intermediary nodes.

The Directed Diffusion routing protocol is a data-centered protocol that requires several paths to reach the destination [INT 00, INT03]. It simplifies the processing of routing processes, such as the suppression of the useless re-diffusion of packets based on a publish/subscribe model with four processing stages: data nomination, propagation of interests and establishment of gradients, data propagation and reinforcement of paths. Through data nomination, an attribute–value scheme is used to describe interests and data by establishing and diffusing requests with an interest and, in response, nodes structure data according to the same scheme as requests, although with new data (Figure 3.3).

3 Dynamic Source Routing.
4 Ad hoc On Demand Distance Vector.

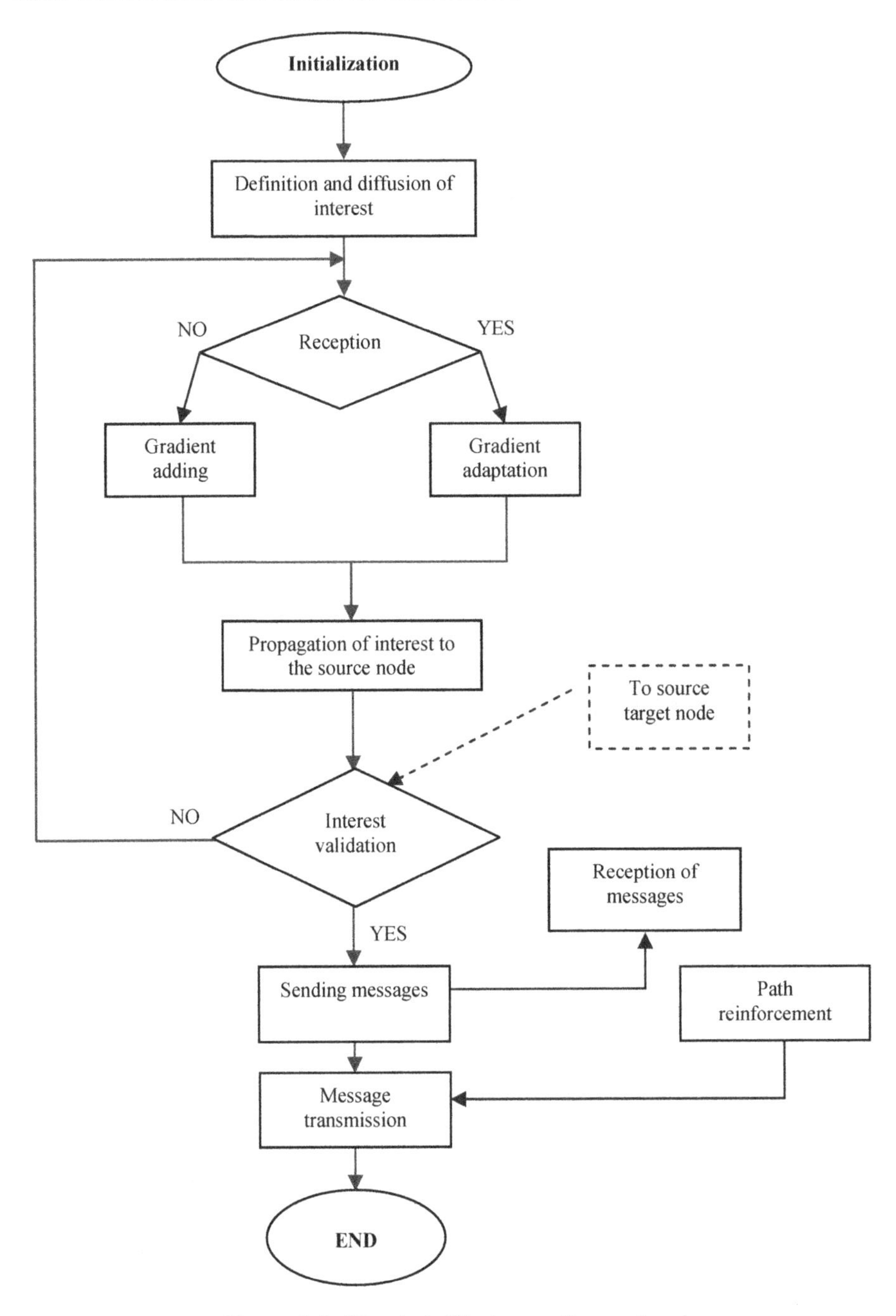

Figure 3.3. *Directed diffusion routing protocol*

When a sink node requires data, it propagates a message of interest with its description and the desired information rate. It explores all paths by flooding the network with messages of interest in order to establish the related gradients that are characterized by a direction modeled by the neighbor transmitting the interest and a range represented by the data rate. Once received, each node examines its cache to confirm the existence of admission in relation to the interest received.

For any given interest, a sensor calculates the highest rate and consequently deducts data. By examining the gradients related to the interest, the node estimates the next hops to the sinks (each with its own rate). If the sensor node receives a message, via its cache, it determines the corresponding interest and, by consulting the list of gradients, it relays the message to its neighbors in accordance with the appropriate rate. In this way, it follows that a neighboring node receiving a message from one of its neighbors, in turn, consults its cache to confirm that it has not received the message previously and to see if there is a relationship with the interests stored. If this is the case, it stores the message and resends it to the next neighbor and, if not, it deletes the message and continues until it reaches the node. If there is no available interest in the cache, then the packet is simply deleted. During exploration, the sink spreads interests at a low rate by establishing gradients in order to strengthen optimal paths. When the sink receives the first exploratory data through any source using several paths, it strengthens the path to the neighboring transmitter by increasing the capture rate. That neighbor adopts the same behavior as the sink until it reaches the target node, i.e. the source, by establishing gradients. The reinforcement must not end with the sink's neighbors, but must eventually propagate to the source. Once the exploration is complete, the information collection begins using a very strict itinerary.

GBR[5] is also a variant of the DD protocol based on gradient [SCH 01]. The idea is to store the number of hops when the interest is broadcast throughout the network. Each node can calculate a node height parameter

5 Gradient-Based Routing.

that corresponds to the minimum number of hops required to reach the base station. The gradient corresponds to the difference between the height of the node and its neighbor, and the message transmitted takes the path where the gradient is highest.

In COUGAR [YAO 02] and ACQUIRE[6] [SAD 03], the WSN is integrated into a data warehouse or a distributed database. Each node incorporates temporarily inaccessible information into a relational table with attributes of the sensor itself or of the data collected. The protocol implements an extended SQL-type interface to integrate modeling clauses of the probabilistic distribution. In order to simplify processing and optimize resource use, the sink node is responsible for generating a request model that allows specific nodes, i.e. CHs, to be selected. These carry out the aggregation and transmission of results. ACQUIRE, meanwhile, is based on a query mechanism for requesting specific data. Each request has several sub-requests that indicate several sensors in order to collect stored data over a specific period. Each node has the possibility to make an active request to the network randomly or by specifying a trajectory to other nodes with a path using a localized updating mechanism to establish a response.

Based on the same principle as directed diffusion, CADR[7] consists of testing sensor nodes before routing the data in the network by maximizing the information gain while also minimizing the latency and bandwidth [CHU 02]. The requests are distributed on the basis of information criteria that allows for the selection of sensors able to acquire data. This occurs by activating only the sensors that are close to a particular event and by dynamically adjusting data routing.

3.2.1.2. *Hierarchical routing*

The hierarchical routing proposed in wired networks is a technique known for the advantages it provides in terms of scalability, communications efficiency and especially the optimization of energy consumption in the case of WSNs [HEI 00]. In accordance with energy levels, sensor nodes have different levels of responsibility and two kinds of sensor node can be identified (Figure 3.4).

---

6 Active Query Forwarding in Sensor Networks.
7 Constrained Anisotropic Diffusion Routing.

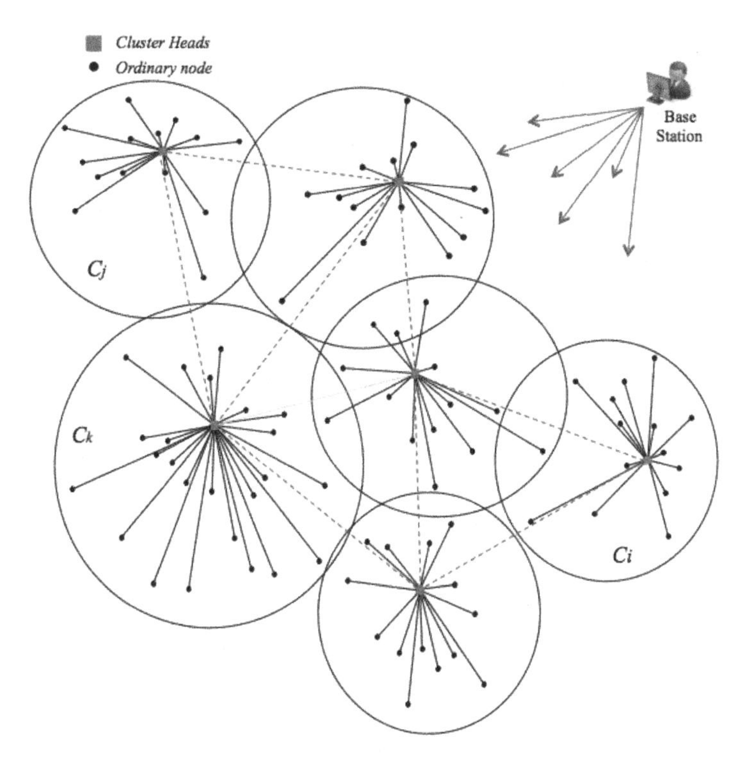

Figure 3.4. *Distributed architecture in a hierarchical topology. For a color version of this figure, see www.iste.co.uk/touati/energymanagement.zip*

There are those with a low level of energy carrying out only sensory measurements, and those with high levels of energy whose objective is to carry out processing and information transmission. This distinction allows it to be made clear that the implementation of clusters represented by the CH master nodes can contribute positively to improving network performance in terms of scalability, durability and energy consumption. Two routing layers are used. The first enables CHs to be selected on the basis of criteria related to its energy level and connectivity with all nodes belonging to the same area and the same geographic position. The aim is to carry out processing, often through the aggregation of data from all nodes within the same category, and to ensure that information is transmitted to the base station or to other CHs as a relay. It is in fact the second layer that handles the actual routing.

LEACH, a highly suitable protocol for WSN monitoring applications or environmental surveillance, is a protocol based on this type of topology, which carries out network classification in clusters [HEI 00]. Nodes are selected randomly as CHs, with the role of communicating with all nodes belonging to related clusters, periodically centralizing the data collected, carrying out processes such as aggregation, and finally directly transmitting the results to the base station if it is accessible or, if not, to other CHs.

Implementation of the protocol firstly covers a network initialization stage where clusters are formed from the preselected CHs, and secondly includes a stage of operating in a stable regime where communication with the base station takes place.

Being based on the TDMA method, the member nodes (MNs) send their data to their CH, in which the data are aggregated then transmitted directly or indirectly to the base station. The inactive MNs have the possibility of entering standby mode to save energy.

Several variants of this protocol have been proposed to improve performance, particularly in the case of dense networks [PAH95, HEI 00]. There is, for example, V-LEACH[8], which undertakes the choice of a CH in the same way as LEACH, as well as a vice CH that takes on the role of the CH in the event of a breakdown or loss of energy [YAN 07]. EACHS[9], another variant proposed in [YAS 09], is based on the implementation of a function that automatically selects CHs, combining LEACH with parameters related to energy consumption. TL-LEACH[10] and M-LEACH[11] are two variants that carry out multi-hop transmission to reach the base station [LIA 05, LOS 05]. As demonstrated in several works [MHA 04], and in contrast to both directed diffusion and SPIN, which are based on optimizing paths and exploiting information in the form of metadata, only LEACH provides better results in optimizing energy consumption (Table 3.1). This is confirmed through the increase and extension of the lifetime of the network.

8 Vice-LEACH.
9 Energy Adaptive Cluster-Head Selection.
10 Two Level-LEACH.
11 Multi-hop LEACH.

Performance	SPIN	LEACH	Directed diffusion
Path optimization	No	No	Yes
Network lifetime	Acceptable	Very acceptable	Acceptable
Resource optimization	Yes	Yes	Yes
Use of metadata	Yes	No	Yes

Table 3.1. *Comparison between SPIN, LEACH and directed diffusion*

TEEN[12] and APTEEN[13], two hierarchical routing protocols intended for critical time applications, are proposed in [MAN 01] and [MAN 02] respectively. The TEEN protocol uses a reactive routing protocol where the nodes are continuously in detection mode with less frequent data transmissions. It builds a group with several branches and various levels until it reaches a data-collecting node. The data-centered aspect is justified by the establishment of two thresholds for captured attributes: a hard threshold which sets the transmission of information from a node to the hierarchical level of the cluster, and a flexible threshold which in turn allows communication only when the attribute has a higher value. As soon as other nodes become CHs, they broadcast other threshold values (Figure 3.5(a)).

The nodes keep their environment under permanent surveillance. Once a parameter of all of the attributes reaches its hard threshold value, the node activates its transmitter module, stores the data detected and then transmits it to its CH. This protocol has proved to be useful as it allows the number of transmissions, and therefore also the energy consumption, to be reduced. However, implementing TEEN is incompatible with applications that collect information periodically. In order to circumvent this, the APTEEN protocol, which is a hybrid that combines both proactive and reactive routing protocols, has been proposed. It enables the modification of periodicity values or the thresholds used in the TEEN protocol according to the needs of the user and the type of application. As shown in Figure 3.5(b), the base station is made up of the CHs that not only distribute the attributes, and the hard and flexible thresholds, as in the TEEN protocol, but also send the delay or schedule transmissions. The periodic broadcast of the information

12 Threshold-sensitive Energy Efficient sensor Network protocol.
13 Adaptive Periodic Threshold-sensitive Energy Efficient sensor Network protocol.

detected is thereby guaranteed. This can also be considered a disadvantage that can slow down processing.

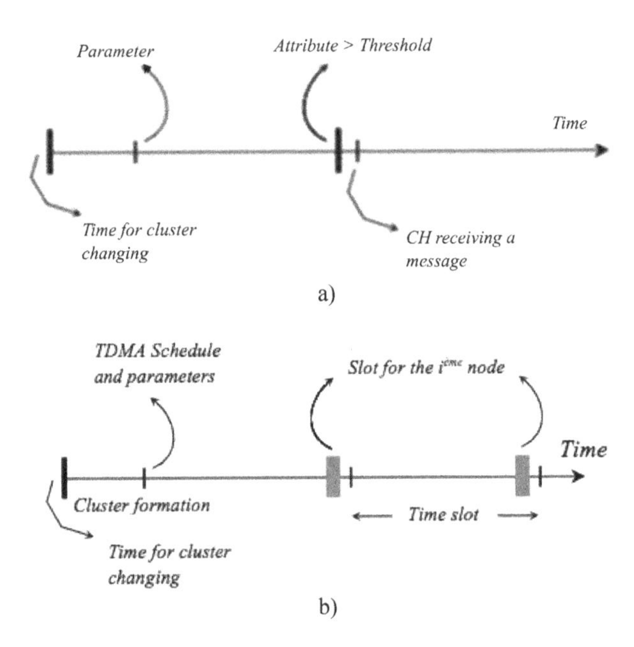

Figure 3.5. *TEEN and APTEEN protocol scenarios*

SOP[14], another protocol with self-organization capabilities, has been developed for heterogeneous architectures with stationary and/or mobile sensor nodes [SUB 00]. The latter survey the environment, and then transmit the data via other fixed nodes acting as routers to the base station. Each node wishing to be part of the network must be able to reach a router, which allows them to be identified as they use the addressing of the router in question. The SOP protocol uses an algorithm based on the implementation of local Markov loops for a random exploration of possible defects in the branches of different trees of the graph, generating, at the end, costs for

14 Self-Organizing Protocol.

maintaining routing tables and preserving the hierarchy. In comparison with the SPIN protocol, SOP consumes less energy during message delivery.

The PEGASIS[15] protocol is another kind of protocol that uses a hierarchical architecture and which brings about improvements in relation to LEACH by organizing all nodes in chains; the nodes alternate at the front of these chains in order to communicate with the base station [ALK 04, SEN 11]. This allows the energy used to be distributed evenly throughout the network. The aim is to randomly position the nodes in the environment so that they can in turn form a chain based on the research algorithm of their closest neighbors. The base station calculates the chain and then broadcasts it throughout the network's nodes. In order to build the chain, the algorithm first considers the sensor node furthest from the base station to guarantee connectivity. When a node disappears, the chain rebuilds itself in the same manner to avoid the dead zone. The capacity to reconfigure the chain gives the protocol robustness and resilience to breakdowns.

The leader of the chain begins the transmission by using a control algorithm passing the token at the end of the chain. The cost of implementing this algorithm is very low as the size of token used is minimal.

More recently, other work has been undertaken to improve the performance of PEGASIS [SEN 11, JUN 07, BAL 11]. H-PEGASIS[16] has been proposed, which carries out simultaneous data transmissions in order to avoid collisions by using approaches that incorporate signal coding and spatial transmissions [LIN 02].

The HEED[17] protocol is an extension of the LEACH protocol with the introduction of a new metric such as residual energy, density or the level of a sensor node used for cluster selection [YOU 04]. The protocol operates within multi-hop networks by adapting the transmission power in the case of inter-cluster communication. The algorithm periodically selects CHs according to two classification parameters: the residual energy of nodes for calculating probability in CH selection, and the intra-cluster communication cost as a function of cluster density or the node degree. The protocol starts

15 Power-Efficient GAthering in Sensor Information Systems.
16 Hierarchical-PEGASIS.
17 Hybrid, Energy-Efficient Distributed Clustering.

with an initialization stage in which the probability of selecting a sensor node as a CH is calculated:

$$CH_{prob} = C_{prob} \cdot \frac{E_r^{(i)}}{E_{Tot}}$$ [3.1]

where the parameters $E_r^{(i)}$, E_{Tot} and C_{Prob} are respectively the residual energy of an *i*th node, the total energy of the network and the optimal number of clusters.

This protocol offers better performance than LEACH in terms of extending the network lifetime, as the selection of CHs is not carried out randomly, which in some cases could cause the rapid disappearance of nodes that are poorly chosen to be CHs.

In the FEED[18] protocol [MEH 11], the network is structured into clusters in which there are ordinary or normal nodes; CHs that guarantee the coordination of intra-cluster and inter-cluster communications, pivot CHs (PCH) with greater capabilities than CH nodes and supervisor nodes (SN). The latter, as their name indicates, ensure supervision of the related clusters able to substitute CH nodes and/or PCH in the event of breakdowns. The PCH nodes cover very important areas in the network and are generally widely used to act as routers (Figure 3.6).

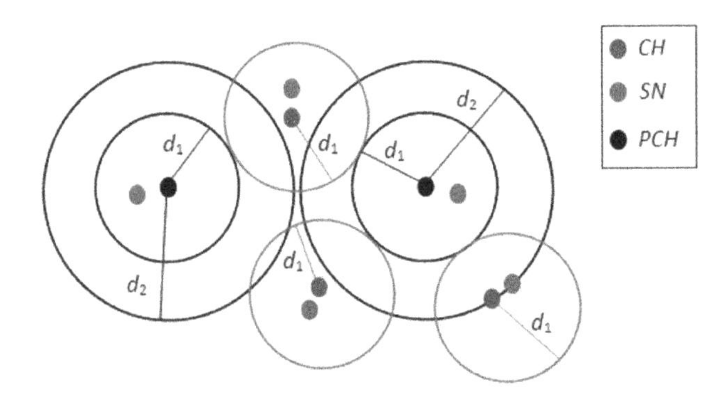

Figure 3.6. *Location of CH, PCH and SN nodes in the network. For a color version of this figure, see www.iste.co.uk/touati/energymanagement.zip*

18 Fault Tolerant, Energy Efficient, Distributed Clustering.

The nodes embed a GPS system that enables them to locate themselves and have knowledge of their geographic coordinates within the network. The implementation has several stages. First, all the nodes must calculate their density factors by distributing their IDs and geographical coordinates to the neighboring nodes, with each node calculating the distances separating them from neighboring nodes upon reception. If the distances calculated are less than a threshold value, the node in question reduces its density factor. If the density calculated is greater than the threshold value and the node is located beyond the borders, the centrality factor is determined in order to establish whether the node can be voluntary or not. The smaller the centrality factor, the more the conditions are favorable to establishing a CH. The ideal is that its value is null, i.e. that the node is located exactly in the center between its neighbors. While the centrality is established, each node calculates a *Sc* score function, such as:

$$Sc = a \, . \, E_n + b \, . \, D_e + c \, . \, C_e \qquad [3.2]$$

where *a*, *b* and *c* are coefficients; E_n, D_e and C_e represent residual energy, the density factor and centrality respectively.

Once the centrality has been calculated, each voluntary node receives a message like: $Sc\_local\_broadcast\_msg\left[myNodeID, C_e, Sc\right]$, before recalculating a second function corresponding to the final score, such as:

$$Zc = d \, . \, Sc - e \, . \, Dist \qquad [3.3]$$

with *d* and *e* respectively being the coefficients of moderating the *Sc* function and of distance.

Each node gives a negative score to the distance factor to force the most distant nodes to become CHs. Each node chooses a voluntary neighboring node with the best *Zc* score as a secondary volunteer and alerts its neighbors. If a volunteer is chosen as a secondary volunteer by several nodes, it will have a higher probability of being selected CH, PCH or SN, and the rules set out in [3.4] must be checked:

$$\begin{cases} d_1 \le d_2 \le \frac{2}{3}.d_1 \\ Min_{distance(PCH_i, PCH_j)} \ge d_2 \\ Min_{distance(CH_i, CH_j)} \ge d_1 \\ Min_{distance(PCH_i, CH_i)} \ge d_1 \end{cases} \quad [3.4]$$

The results of simulations show that the FEED protocol use enables a significant improvement in classification in particular to extend the lifetime of the network in comparison with LEACH and HEED.

3.2.1.3. *Location-based routing*

In this type of routing, the sensor nodes are identified by their location within the network. The distances between neighboring nodes can be estimated in two ways: either through the exchange of RSSI information and use of empirical models formed from environmental and energy parameters [SAV 01, BUL 00, CAP 01], or directly by using the satellite system if the sensors embed low power GPS receivers [XU 01]. All sensor nodes in accordance with their operational activities within the network adopt the so-called active or sleep modes.

Initially intended for mobile and ad hoc networks, the GAF[19] protocol enables energy consumption to be optimized with an acceptable level of quality of service (packet loss, message latency) [XU 01]. The detection area is divided into small adjacent and equivalent virtual grids, with each having a single node (i.e. leader) that can be activated at any given moment. This involves the coordination and diffusion of messages across all nodes in each grid in order to select which will be activated or deactivated (sleep mode) for a period of time. The leader periodically rebroadcasts a discovery message in order to allow other nodes in the grid to be activated in turn. As shown in Figure 3.7, the status transition diagram shows three states related to a sensor node: search-discovery, active and sleep. When a node puts itself in sleep

19 Geographical Adaptive Fidelity.

mode, it switches off its radio to save energy. In search-discovery mode, discovery messages are sent between sensor nodes in the same grid in order to inquire about their neighbors. These discovery messages are also periodically broadcast when the node is active to inform other sensors of that fact. The time spent in each of these modes can be adjusted to the necessary application in accordance with the needs and mobility of the sensor.

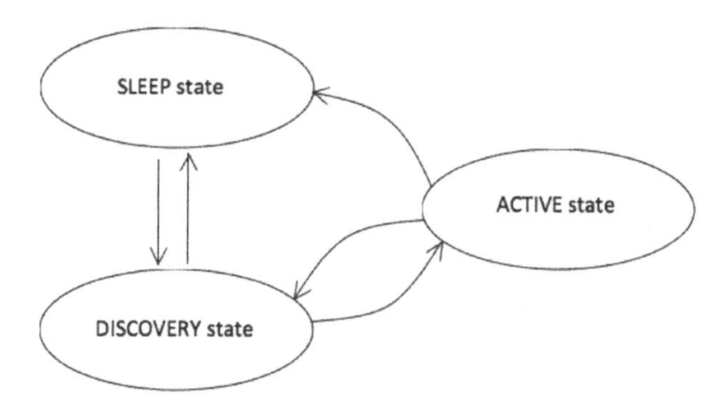

Figure 3.7. *State transition diagram*

Although it is based on location, GAF can also be considered a hierarchical protocol, where clusters are formed on the basis of geographic location. For each area of a particular grid, a typical sensor node acts as a leader to ensure that data is transmitted to other nodes. On the other hand, unlike in hierarchical protocols, the leader node cannot guarantee the aggregation and data fusion functions. Simulation results have shown that GAF is efficient with regard to latency, packet loss and increasing the lifetime of the network by optimizing energy. However, the network can be underused overall given that the protocol is based on communication between adjacent grids using at least half of the radio range.

Staying with the same family of protocols that exploit the position of sensor nodes, in GeRaF[20] [CAS 05, ZOR 03a, ZOR 03b] the nodes initially put themselves in listening mode before changing periodically according to operating cycles, alternating between active and sleep modes. When a node wishes to transfer data, it changes to active mode and sends a packet

20 Geographic Random Forwarding.

containing its own location and the location of its target. The latter responds with a message of acknowledgement in order to begin transmission. The area close to the target is divided into regions which each have their own priority. The closer the region is to the destination, the higher its priority. After the packet has been diffused, the regions with nodes very close to the destination are called on to transfer and relay information. If all the nodes in a region are in sleep mode, a transit attempt is made through another high priority region, and so on.

The GEAR[21] protocol developed in [YU 01] exploits geographic information by diffusing requests into appropriate regions, in view of the geographic attributes that the requests have. In order to ensure that information is routed, the protocol uses heuristics in order to select neighboring nodes through geographic information and related energy levels. The idea is to limit the search areas by broadcasting interests, as in the case of directed diffusion, which allows energy to be saved more effectively. Each sensor node saves the estimated cost, which combines the residual energy and the distance separating a node from the target, and the cost related to training the routes to reach the targets through their neighbors. The training cost corresponds to improving the estimated cost for routing around an isolated node or hole without enough close neighbors, thereby allowing it to reach the target region. If there are no holes, the estimated cost will be identical to the training cost. The latter propagates back in a single hop each time that a packet reaches its destination in order that the route configuration can be adapted when the next packet is sent. In comparison to GPSR[22] [KAR 00], which addresses issues of node isolation by using the concept of two-dimensional graphs, the GEAR protocol also ensures a better rate of data packet delivery. In contrast to GAF, GEAR cannot ensure the aggregation or data fusion functions and scalability is often not guaranteed during packet transmission [ROY 10]. Implementation of the routing mechanism often overcharges the network and therefore directly influences energy efficiency. A major difference between the two protocols concerns the use of the data delivery model. The GAF protocol follows a data delivery model based on a virtual grid with master–slave transmission, while in GEAR protocol the model is based on the principle of requests to send information.

21 Geographic and Energy Aware Routing.
22 Greedy Perimeter Stateless Routing.

Initially designed for MANET networks, the SPAN protocol ensures better coordination between sensor nodes to optimize energy [CHE 02]. It carries out the selection of coordinator nodes according to their positions in the network in order to form a main sub-network for routing data. A coordinator node is chosen when two neighbors of a non-coordinator node cannot communicate either directly or through one or two coordinator nodes. In combination with a geographic transmission protocol, the rule for choosing in the SPAN protocol requires each sensor node to announce its status to its neighbors, i.e. coordinators or non-coordinators.

The TBF[23] protocol implemented in dense networks also requires the use of a coordinated GPS system allowing the sensor nodes to position themselves within the environment and calculate the distances separating them from neighboring nodes [NAT 03]. The source specifies the trajectory to be followed in the packets sent but does not explicitly indicate the route hop-by-hop. On the basis of the location information of neighboring nodes, the sensor node emitting the data packets decides on the next hop to be made, which must be as close as possible to the trajectory defined by the source. In order to increase the network's reliability and capability, it is possible to implement a multi-path routing in which an alternative trajectory can be used. The protocol can implement several functions such as flooding, path discovery and network management.

In the same family as TBF, the auto-reconfigurable MECN[24] protocol is able to guarantee network connectivity, regardless of whether stationary or mobile sensors are used [ROD 99]. It allows a minimum-energy topology to be designed by building optimal trees in the form of chains to the destination sink node. This design is based on the positions of nodes in the operational environment and comes into being through a two-stage implementation process; namely, the establishment of the graph and the cost distribution. While the protocol is able to reconfigure itself, and can therefore tolerate defects and breakdowns (in the case of mobile networks), it suffers from battery exhaustion when dealing with sensor network applications.

An improved version of the MECN protocol is proposed in [LI 01b]. It concerns the SMECN[25] protocol in which a minimal graph is distinguished

23 Trajectory-Based Forwarding.
24 Minimum Energy Communication Network.
25 Small Minimum-Energy Communication Network.

according to the minimal energy. In other words, for each pair of sensors in a graph, there is a minimal path with very low costs in terms of energy consumption across all possible paths. Each node discovers its immediate neighbors by sending a neighbor discovery message using its initial power. It then confirms whether the theoretical group of immediate neighbors corresponds to a subset of all sensors that responded to the discovery message. If this is the case, the sensor node will use its initial power to transmit the message, and, if not, it increases this power and retransmits its discovery message to its neighbor.

3.2.2. *Routing based on protocol operation*

Other routing protocols belonging to the family of protocols based on location have been proposed, such as GOAFR[26], GEDIR[27] and MFR[28] [STO 99, KUH 03]. In this protocol family, implementation depends greatly on the types of application. Among the techniques used are negotiation, the establishment of requests, the verification of data consistency, multi-path diffusion and, lastly, QoS.

3.2.2.1. *Multi-path routing*

Fault tolerance is one of the main features that a routing protocol must have in order to ensure the continuous operation of the network. A break in the path due to a sensor malfunction, i.e. physical defect or lack of energy, must not in any case cause a problem in transmission between the source and the destination. The communication protocols based on multi-path routing can be remedied by proposing alternative paths in order to improve network performance in terms of robustness, overloading and longevity. Maintenance of all alternative paths is highly important and guaranteed through periodic control by conveying so-called control messages.

Various works using different criteria, such as residual energy, have been proposed and developed to this effect. In [CHA 04], a protocol is implemented to choose the best branch of a tree that information can follow while taking account of the value of the pre-established energy threshold. By using this principle, the energy resources of nodes belonging to other initial

26 Greedy Other Adaptive Face Routing.
27 Geographic Distance Routing.
28 Most Forward within Radius.

branches will not run out straight away. However, in such methods the costs related to changing an itinerary during transmission cannot be quantified.

In [RAH 02], an algorithm randomly calculates a set of optimal paths according to a probability dependent on changes to energy consumption at the level of each branch. Knowing that a branch with significant residual energy can also use further energy to route information, it must therefore have a compromise between minimizing consumption and residual energy. For this purpose, it may be useful to have a routing procedure that allows switching between the different branches of a tree. This is used in the protocol proposed in [LI 01a]. When the residual energy throughout a branch reaches a minimum threshold, the algorithm redirects the information along another branch that consumes less energy.

In [DUL 03], the algorithm is able to improve the reliability of networks, particularly those deployed in hostile environments, by broadcasting information across several branches in order to increase the probability of conveying and delivering data. However, the traffic can significantly increase, causing the network to overload and therefore requiring a compromise between traffic density and network reliability. In order to do this, a redundancy function depending on the degree of data propagation and the probability of failure of routing on the available paths has been proposed. The idea is to structure data packets initially into a set of sub-packets able to be transmitted across several different paths. This is an interesting concept as, even if some sub-packets are lost, the original message can always be reconstructed at the target level. Protocols based on directed diffusion [INT 00] also seem to be good candidates for strong multi-path routing.

3.2.2.2. *Request-based routing*

As in the case of directed diffusion [INT 00], the destination nodes propagate a message written in a high-level natural language into the network searching for a piece of data, i.e. in the case of a detection task. The node with the information relevant to the requests sends the packet of information to the node that initiated the request. The base station sends messages of interest throughout the entire network, before the gradients from the source node to the base station are established. Once the source has data with a specific interest, it sends its data across the network in accordance with the gradient of interest by carrying out data aggregation, i.e. eliminating duplications. This allows energy to be optimized.

In this type of routing, the RR[29] routing protocol [BRA 02] is a perfect example as it uses a range of active agents to create paths targeting encountered events. Each agent has a lifetime that allows it to ensure a certain number of hops, before disappearing from the network. Each time it encounters a new event, it creates a status on the path taken by recording both the event and the itinerary followed that is able to lead to the event if necessary. When the agents take very short paths, they update by subsequently adapting their routing table by re-optimizing routes and other events if they exist. In a probabilistic manner, each node can also generate an agent with a table of events that is synchronized with another sensor node. In this type of RR routing, a node can only generate a request if it has knowledge of the path essential to reaching the event. If no path is possible, it transmits a request randomly before putting itself in standby mode during the interval, after which the node floods the network if it has not received any response from the destination.

3.2.2.3. *Routing based on negotiation*

This type of routing has been developed to resolve implosion and overlapping between data when the network is flooded with messages. Each node can receive several copies of the same message and can therefore be susceptible to consuming much more energy during transmission and processing. Consequently, the main idea of negotiation in WSNs is to use high-level data descriptors in order to put an end to the duplication of information and avoid data redundancy. The operating principle of negotiation is based on the use of three types of messages: REQ messages sent by sensor nodes wishing to receive data; DATA messages containing data with a metadata heading; and ADV messages that are sent as announcements when a node has data to be shared within the network. ADV and REQ messages are smaller than DATA messages, and do not contain metadata. This is because, in a network where the cost of transmitting and receiving messages is largely determined by the size of the message, ADV and REQ messages consume less energy than DATA ones.

The SPIN-PP, SPIN-EC, SPIN-BC and SPIN-RL protocols are part of the SPIN family in which routing is based on negotiation [YE 01, HEI 99]. The SPIN-PP and SPIN-EC protocols are intended for peer-to-peer networks for which data packets are never deleted and which never have delays caused by

29 Rumor Routing.

node sensor traffic. SPIN-BC and SPIN-RL, meanwhile, are related to diffusion-based networks.

3.2.2.4. *Routing based on QoS*

Implementation of a routing protocol based on QoS requires the consideration of criteria or metrics such as energy, latency, bandwidth and so on. SAR[30] is one of the first routing protocols to take QoS into account [ALK 04]. As regards metrics, this protocol is based on the energy consumed, the QoS on each path and the priority level of packets. It implements a multi-path approach with methods of restoring targeted paths in order to prevent flawed itineraries, and therefore be fault tolerant.

The creation of multiple paths through a source node involves the construction of a routing tree from the source to various destinations. Once this process has been achieved, each node becomes part of the structure of the tree that has been built. It calculates an average metric balanced by a coefficient for each priority level of the packet to be routed. The presence of possible defects that could cause a change in topology is considered directly by calculating new paths, requiring a readjustment of the different routing tables and the states of each node. This readjustment can cause concerns with regard to energy consumption due to overloading, which is the case in dense networks in particular.

Within the same family, there is another protocol, SPEED, in which each node must have information on its neighbors allowing it to ensure network performance in real time from beginning to end [HE 03]. Each packet transferred at high speed allows a data application to estimate the routing time from start to finish by dividing the distance source-BS by the pre-established speed. Moreover, SPEED prevents network congestion as it is based on informative messages sent by congested nodes to their predecessor nodes in order to select alternative paths. In order to do this, it has an integrated SGNFG[31] routing module that estimates the time limits at the level of each node by calculating the time spent following receipt of an acknowledgement message ACK from a neighboring node as a response to a transmitted data packet. A comparison of the different time periods allows a node with sufficient packet transmission speed requirements to be selected.

---

30 Sequential Assignment Routing.
31 Stateless Geographic Non-Deterministic Forwarding.

Compared to the DSR[32] and AODV[33] protocols, the SPEED protocol offers the best guarantees in terms of routing delays and loss rate. The energy consumed during message transmission is lower due to the simplicity of the protocol. However, SPEED does not take into account other energy metrics in the routing process and, consequently, it is very difficult to realistically compare it to other kinds of protocol. The holes caused by, for example, defects in several nodes in any geographic area are not considered in themselves, but as temporary congestion phenomena.

3.2.2.5. *Routing based on data consistency*

Generally, in a WSN, the nodes carry out processes and communicate with each other to share various kinds of information, leading the routing process to take both the consistency and inconsistency of data into consideration [ALK 04]. In the case of consistent data, the data packet is conveyed to aggregator nodes once the processes have been established, such as the deletion of redundant data. This is efficient routing in terms of energy optimization.

On the other hand, in the case of inconsistent data, the nodes process raw data locally before it is routed to other nodes for further processing, i.e. aggregation. In this way, the processing encompasses three stages. The first stage is target detection, the second one is data collection and the third on is pre-processing. When a node needs to cooperate, it enters a second stage of declaring membership, where it declares its intention to its neighbors. This must be carried out as far as possible in a way that each node has local knowledge of the neighborhood and the network topology. In the third stage, a central node must be chosen in order to carry out more efficient processing and, in order to do this, it must have sufficient energy and significant calculation capabilities.

3.3. Critical analysis

In this chapter, we have concisely illustrated the issues and challenges in WSNs, before exploring routing techniques developed in recent years by drawing attention to factors improving and/or worsening network performances and functioning.

32 Dynamic Source Routing.
33 Ad hoc On Demand Distance Vector.

Increasing the performance of a WSN in terms of longevity, connectivity and robustness requires that essential factors such as the energy consumed and the calculation and memory resources be taken into consideration. It is therefore necessary to find alternatives, particularly software alternatives, as once a network is deployed it cannot withstand repeated human interventions for each malfunction it might have, i.e. battery failure, environmental influence and so on. Moreover, these malfunctions can cause significant topological changes, requiring a complete reorganization of the network in order to ensure the exchange and conveyance of data. Proposing suitable solutions based on routing can bring about important advances in improving the rate of information exchanges by considering topological changes and stabilizing the network operation over time by making optimal use of energy consumption. A data-centered routing intended primarily for networks with flat topologies and which are known for their high tolerance of breakdowns during route construction, with very low maintenance costs, can be considered as a solution. Nonetheless, it uses a significant number of control messages to ensure that the network functions well without actually improving scalability and therefore degrading performance in some cases. Moreover, the nodes located close to the base station are often required and if their loss is accelerated it can lead to an overconsumption of energy, reducing the lifetime of the network. On the other hand, the hierarchical routing known for the advantages it offers in scalability and transmission efficiency can provide solutions in optimizing energy consumption, particularly for dense heterogeneous networks. During the deployment stage, CHs, which are more powerful in terms of energy, bandwidth and memory and which function differently to ordinary nodes, are formed and then chosen. They allow the number of messages conveyed to the target to be optimized through the aggregation of data from different sources, considerably reducing energy consumption. Location-based routing identifies the location of nodes in the network by estimating the distances between neighboring nodes, either by exchanging RSSI information and using empirical models made up of environmental and energy parameters, or directly using a satellite system and, in this case, the sensors embed low power GPS receivers.

Furthermore, the implementation of a routing protocol must take into account and, if possible, optimize all sources of energy consumption, which are numerous and do not only concern the transmission and reception of data [HAL 09]. These sources include processing carried out by the processor, the detection of environmental information, transitions between the different states of a sensor such as the standby and active modes, reading and writing memory, actuators and, of course, the creation of clusters in the case of hierarchical architectures. Each of these sources consumes a different amount of energy, and the percentages of energy loss for a WSN with 100 nodes shown in Figure 3.8 are proof of this.

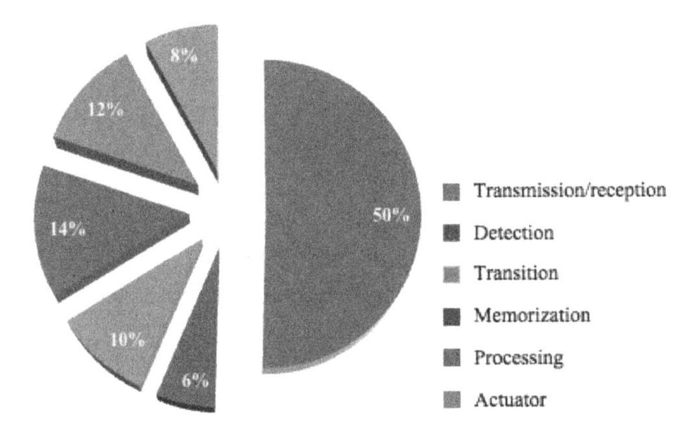

Figure 3.8. *Energy consumption rate in a WSN. For a color version of this figure, see www.iste.co.uk/touati/energymanagement.zip*

Over 50% of energy losses are in fact caused by the radio; that is, the transmission and receipt of data. The greater the distances, the more significant the amount of energy required to route information. Depending on the WSN applications, the energy sources can be taken into consideration or not. For instance, an application for monitoring forest fires does not consider energy used on actuators, or even on storing or saving data, and, in this case, over 70% of the energy used goes on radio transmission.

Maximizing and estimating the lifetime of a WSN require the implementation of an energy consumption model that suits the application in question. Several pieces of research have examined this context and proposed more or less specific models that specify certain sources of energy consumption. The Heinzelman model [HEI 02], for example, does not take into account energy models related to processing or radio transmission. Miller [MIL 05] considers the status transitions of sensor nodes. The Zhu model [ZHU 03] incorporates energy spent on detection in the energy used on processing and communications. Models related to detection, memory and actuators are rarely considered. Proposing a routing solution must therefore occur primarily in accordance with the context and the type of application considered by identifying the sources of energy consumption and, above all, the parameters to be taken into account. It is thus important to not choose a specific routing procedure at the expense of another, but to draw from the entire range of existing solutions.

In the chapter to follow, we will look at dense networks, particularly those structured hierarchically, and with the objective of improving performance in energy consumption and information routing delays on the basis of consumption models related to processing, radio and cluster formation.

4

Adaptive Routing for Large-Scale WSNs

In WSNs, the efficient use of information largely depends not only on the processing and exploitation of data, but also on the methods that allow it to be routed. It is therefore necessary to consider the operational and/or structural constraints, namely the intrinsic characteristics of sensors (energy consumption, calculation and memory) and environments (network topology, lack of infrastructure, loss of nodes). Otherwise, communications between different sensor nodes must obey a routing protocol determined in advance in accordance with the type of application and the network architecture, be it flat or hierarchical. A suitable choice of protocol must not only allow the fluidity of information but also the optimization of energy consumption and the use of resources (calculation time and storage capacity). This is an ongoing challenge, particularly in the case of dense networks.

4.1. Introduction

A highly suitable protocol for WSN monitoring applications or environmental observation, LEACH[1] is a protocol based on hierarchical platforms. It allows the network to be structured into clusters with the aim of increasing network coverage, and extending its longevity while optimizing energy consumption [HEI 00]. Sensor nodes are chosen randomly like Cluster Head (CH) to represent different clusters and with the role of communicating with all nodes belonging to their own clusters, periodically centralizing the data collected, and carrying out processing, i.e. aggregation,

1 Low-energy adaptive clustering hierarchy.

before transmitting it directly to the base station, if possible, or to other CHs, if not.

As illustrated in Figure 4.1, implementation of the protocol first requires a network initialization stage where clusters are formed from pre-selected CHs, and, secondly, a stage of stable operation where communication with the base station takes place.

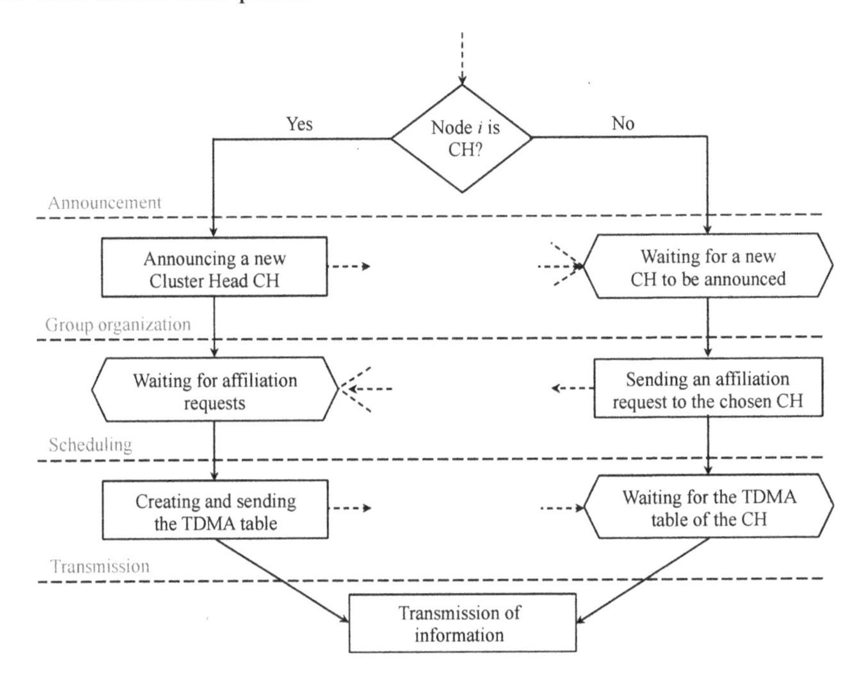

Figure 4.1. *Implementation stages of the LEACH protocol*

During the announcement and cluster creation stage, the base station announces the beginning of a new round. With a pre-established percentage of CHs to be selected (5%–15%), the nodes themselves choose which will be CHs according to a selection probability determined as follows:

$$P_{rob}(n) = \begin{cases} \dfrac{P}{1 - P \, mod\left(\dfrac{1}{P}\right)} & if\ n \in G \\ 0 & if\ not \end{cases} \qquad [4.1]$$

The P and r parameters respectively represent the percentage of nodes wishing to be CHs and the iteration or current round, with G being all nodes that are not CHs in the last $\left(\frac{1}{P}\right)$ iterations, and n being a random value formed in the interval [0,1] attributed to a node. If the latter is less than a threshold, $T(n)$, the node declares itself as a CH and then informs its neighbors of its choice.

An advertisement message containing its identifier as a CH is thus broadcast via a MAC CSMA/CA protocol that allows potential collisions and interferences between different adjacent CHs to be avoided. The nodes decide to belong to CHs by taking into account the amplitude of the signal received. CHs that broadcast signals of a higher amplitude have a greater chance of being chosen than other CHs. All things being equal, the choice of which CH to belong to is made randomly. Thus, a packet of affiliation is sent to the chosen CH, which sends a message of acknowledgement as confirmation.

On the basis of the task scheduling method, it implements a MAC TDMA protocol and assigns each of the NMs from its cluster a time slot during which the node can communicate information. The set of slots assigned to nodes in a group forms a frame, whose lifetime depends on the number of nodes in the cluster [YIC 08]. Figure 4.2 shows the different stages of CH formation.

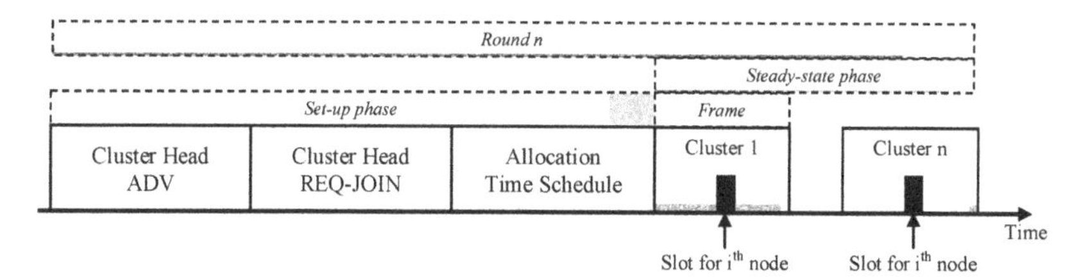

Figure 4.2. *Stages of CH formation*

The transmission stage allows decisions to be made on the performance and robustness of the routing algorithm. Being based on the TDMA method, the NMs communicate their data to their CHs, which are aggregated and then transmitted, directly or indirectly, to the base station. The inactive NMs are able to put themselves in standby mode to save energy.

In contrast to several works such as those using directed diffusion and/or SPIN protocols [YE 01], which are based on path optimization and modeling information in the form of metadata, the LEACH protocol is the only one to provide better results in terms of optimizing energy consumption and therefore increasing the lifetime of the network [ALK 04]. This is one of the main motivations that has led to the choice of this subject as the basis of our work in this book.

In this protocol, data processing at the level of each cluster is carried out locally and the role of each CH is to coordinate exchanges with other member nodes, NMs. The network has the ability to reorganize itself automatically during the CH selection stage. Each node has the potential to be chosen as a CH and vice versa, as each CH can once again become a simple MN belonging to a cluster.

The choice of a CH is based on energy criteria. The greater the amount of energy a node has, the more likely it is to become a CH. Once the CH has been chosen, it communicates with all MNs according to a TDMA MAC layer access protocol. Among other functions, it carries out the aggregation of data coming from MNs, thereby optimizing the processing time. In contrast to MNs, CHs are continuously active while they communicate, whether they communicate with the range of nodes belonging to their clusters, or whether they exchange data with the base station. This process greatly reduces the residual energy of different CHs, thereby also reducing their operational capabilities and the performance of the network. In the case of large-scale networks, this protocol no longer offers a suitable solution to increasing performance in terms of resource optimization, energy consumption and delays for processing/routing data. In fact, during the stage of announcing and creating clusters (*Setup phase*), the selection and distribution of CHs in the network is carried out randomly according to a round-robin management policy (guaranteeing uniform energy consumption across all nodes), which, in some cases, leads to a high concentration at the area level in comparison to others and, consequently, having completely isolated nodes (without CHs) and therefore disappearing from the network. The choice of CHs does not assume any restriction on either distribution or their level of energy. Moreover, situations can occur where all random numbers generated by sensor nodes are less than the probability generated by the base station and, in such cases, no CH is chosen. The base station must quite simply take back control by broadcasting a new probability via an announcement message from a new round, causing an unnecessary overconsumption of energy and a waste of time.

4.2. Adaptive routing mechanisms

The main objective of routing mechanisms is to provide significant improvements in the performance of standard protocols such as LEACH. Its development and implementation includes the establishment of satisfaction criteria, namely, energy consumption and delays for processing and routing information to a destination, regardless of the position in the base station or adjacent CHs. Indeed, in contrast to LEACH, which assumes that all nodes have sufficient energy to communicate directly (one hop) with the BS, an adaptive protocol can, by inheriting information [CHE 06, KHA 10, KUM 11], do it in several hops, limiting energy consumption, particularly for the most distant nodes, and therefore also their isolation and eventual disappearance. The mechanism also allows CHs to be distributed evenly across the network and, in particular, limits the rotation number of the CH when the base station distributes new roles.

As illustrated in Figure 4.3, an adaptive routing approach relies on a clusterization mechanism and the routing of information that revolves around three main stages: first, a deployment and network initialization stage in which sensor nodes with well-defined features specific to the type of application are deployed randomly in operational space.

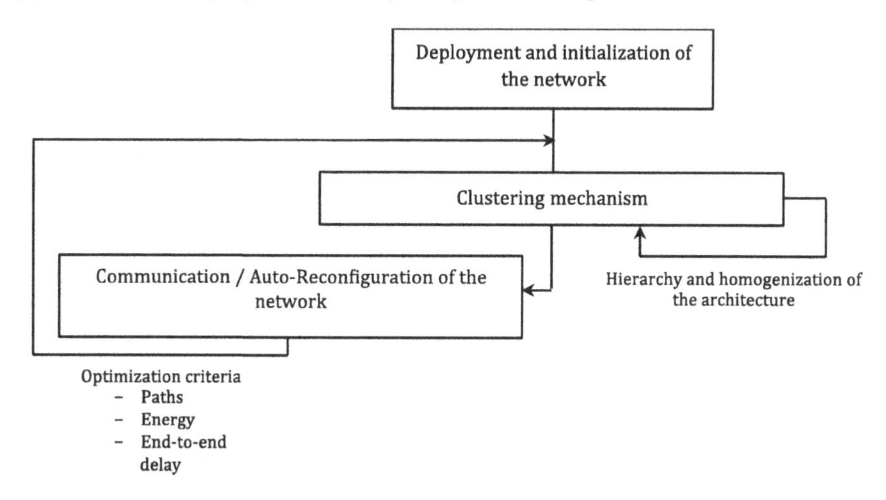

Figure 4.3. *Summary of the basic elements of an HHRP mechanism*

The cluster construction stage, which is the heart of the mechanism, consists of structuring the network into a set of islets in which master nodes

(CH) guarantee the coordination of all member nodes and processing. Master nodes are able to communicate directly (single hop) with the base station or in several hops by passing by other master nodes acting as relays. The hierarchical architecture of the network shown in Figure 4.4 is a good example. We can clearly see a network structured in islets or clusters with each having its own CH receiving data from all of its MNs or communicating directly or indirectly with the base station. The choice of CHs and their MNs must obey pre-established selection criteria in accordance with the application and the network density. In practice, the number of CHs varies between 5% and 15% of the total number of nodes. A network of 20 nodes cannot contain 15% of CHs, as it is low density, or a network of 500 nodes can have 5% of CHs. A poor choice in this number leads inevitably to the rapid extinction of the network. It is therefore necessary to estimate the number of CHs upon initialization. A number of works of research have been undertaken to this end [AOU 12a, AOU 12b], but, in general, simulations have shown that, whatever the dimension of the network, the optimal number of CHs at initialization is in the range of 10%.

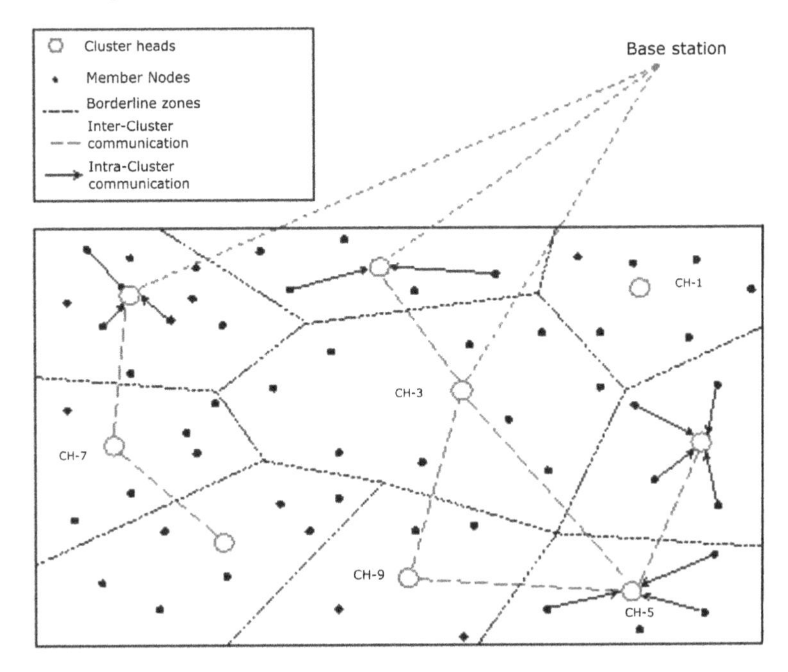

Figure 4.4. *Hierarchical network architecture. For a color version of the figure, see www.iste.co.uk/touati/energymanagement.zip*

The communication stage corresponds to data transmission between nodes according to pre-established routing protocols. The objective is to select, by optimizing criteria, the best path for routing data to the destination. For instance, the CH-5 master node has several possibilities for routing information to the base station. It can use several sequences, such as CH-5-CH-9-CH-3-SB and CH-5-CH-3-SB. The question is in which path to choose to reach the base station by optimizing the use of resources, the energy consumption and conveying delays, and so on.

The idea is therefore to determine one or several performance functions, which allow all parameters to be taken into consideration, such as the distances covered from the source to the destination, the end-to-end delays, signal strength, energy consumption and/or number of hops. For this purpose, it is essential to determine the structure and topology of the WSN to be deployed.

4.2.1. *Network structure and topology*

The implementation of the mechanism is based on a hierarchical architecture, made up of a set of clusters and/or areas. Each cluster is formed from a set of nodes called member nodes, MNs, with a representative node acting as the coordinator and known as the Cluster Head or CH. As in LEACH, it allows processing to be carried out and information to be relayed directly between cluster nodes and the base station or indirectly via other CH nodes. This depends on the range of the signal and the coverage area. The further the nodes are from the base station, the more they use the services of other nodes to reach the destination. Communication is guaranteed through two hierarchical levels: intra-cluster and inter-cluster (Figure 4.5).

Intra-cluster communications, i.e. Intra-Hs communication, concern exchanges made in a given cluster between the set of MNs and their corresponding CH. This can affect, for example, messages of belonging, the diffusion of states, the attribution of temporary slots or, very simply, effective data transmissions from MNs to CHs.

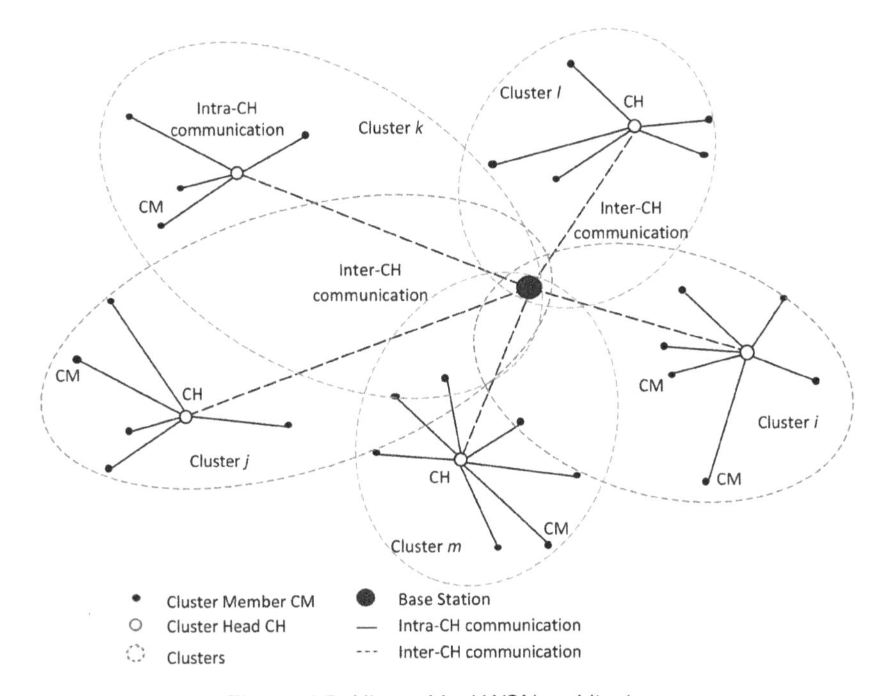

Figure 4.5. *Hierarchical WSN architecture*

On the other hand, inter-cluster communications, i.e. Inter-Hs communication, include information exchanges between different CHs, or between CHs and the base station. A CH located at an unreachable distance from the base station and wishing to transmit a message to it has the possibility of passing through several neighboring CHs. It must therefore take into consideration the different restrictions imposed by the application in order to determine the optimal path. It will reach the base station through one or multiple hops (multi-hop).

The advantage is that if a node does not have enough energy to reach the base station, it can route its data via other CH nodes so that the data is relayed to its final destination. It should be noted that the use of multiple hops optimizes energy consumption, as the transmission distances are shorter. Several CHs are required, thereby enabling a uniformity and homogenization in the energy spent across the network. The drawback of having multiple hops is that it can affect the information routing delay, as it must cross several sections to arrive. The aim is therefore to determine the best choices to be made to transmit the data from a source to a destination by reconciling several parameters, namely, energy consumption, signal strength, end-to-end delays for routing information and the number of hops.

The hierarchical architecture presented above is based on the use of an energy model that takes into account the different states or operational modes of a sensor node in the network. In transmission/reception modes, the sensor nodes use greater amounts of energy to transmit or receive messages structured as bytes. The larger the number of bytes, the more important the messages are and the more the energy spent as a result. In order to optimize the number of bytes to be processed, we use the principle of data aggregation on the basis of mathematical operations; the average, in our case.

Once the radio is activated, the sensor node finds itself in an active state without actually transmitting or receiving data, thereby causing an additional unnecessary waste of energy. This is the case of CHs that are continually awaiting a message. In order to avoid this, the sensor node can sleep by disconnecting its radio. The transition from sleep mode to active mode requires several pieces of software and components to be restarted, involving a greater consumption of energy than if the sensor node stayed in active mode. The energy management can be guaranteed through a protocol in the MAC sub-layer.

In the section to follow, we will present the energy model used in the framework of the proposed routing approach.

4.2.2. *Energy model*

As illustrated in Figure 4.6, the energy process under consideration uses a consumption model that uniquely covers both the data transmission and reception modes [HEI 00]. This is a widely used universal model that has proven its efficiency.

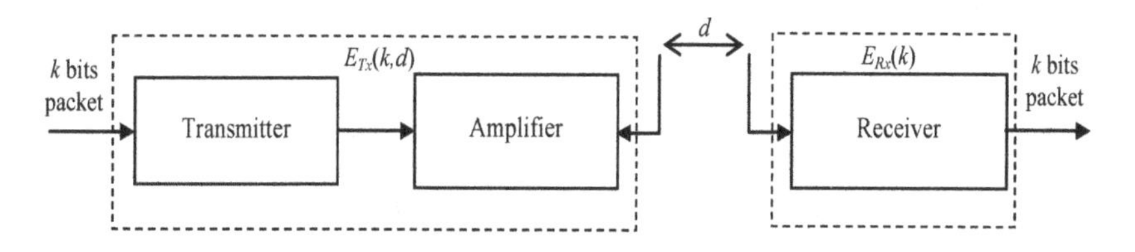

Figure 4.6. *Energy consumption model*

In this way, in order to transmit a coded message in *k* bits over distance *d*, the transmitter consumes:

$$E_{T_x}(k,d) = E_{TX_{elec}}(k) + E_{Tx_{amp}}(k,d) \qquad [4.2]$$

with:

$$E_{TX_{elec}}(k) = kE_{elec} \qquad [4.3]$$

and:

$$E_{TX_{amp}}(k,d) = \begin{cases} k\varepsilon_{fs}d^2 & si\, d < d_{crossover} \\ k\varepsilon_{tr}d^4 & si\, d \geq d_{crossover} \end{cases} \qquad [4.4]$$

During radio communications, the energy consumed at the reception level is calculated as follows:

$$E_{R_x}(k) = E_{RX\_elec}(k) = kE_{elec} \qquad [4.5]$$

The E_{elec} and $E_{Tx_{amp}}$ parameters represent the electronic transmission/reception energy and the amplification energy respectively, with ε_{amp} being the amplification factor and $d_{crossover}$ the distance limit for which the transmission factors change in value.

Once they are selected, CHs remain continuously active, guaranteeing intra-CH and/or extra-CH communication. Being based on the radio model described in [4.2] and [4.5], the average energy consumed by each CH can be calculated as follows:

$$E_{moy} = p_r \left(E_{T_x}(k,d) + E_{R_x}\left(\frac{T_{inter}}{T} - k \right) \right) + (1 - p_r)\left(E_{R_x}\left(\frac{T_{inter}}{T} \right) + E_{R_x}\left(\frac{T_{intra}}{T} \right) \right) \qquad [4.6]$$

where p_r is the likelihood of each node having data coded in *k* bits to send in any given cycle. The *T* parameter represents the time taken to send a byte of data, with T_{inter} and T_{intra} respectively being the communication time between different CHs, and the communication time between CHs and the member nodes during a given round.

In the first part of [4.6], for a probability p_r corresponding to an inter-CH communication stage, all CHs transmit information to the base station, consuming energy equivalent to $E_{T_x}(k,d)$. The time remains $\left(\frac{T_{inter}}{T}-k\right)$, corresponding to the listening time of the CH, of which the energy consumed is $E_{R_x}\left(\frac{T_{inter}}{T}-k\right)$.

In the second part of [4.6] corresponding to a probability of $(1-p_r)$, the CH does not transmit any data to the base station. It spends all of its inter-CH communication time in listening mode, consuming energy equivalent to $E_{R_x}\left(\frac{T_{inter}}{T}\right)$. During the intra-CH communication stage, the CH enters reception mode, consuming energy equivalent to $E_{R_x}\left(\frac{T_{intra}}{T}\right)$.

In the following chapter, we present an adaptive routing solution that uses inheritance and which is based on a clusterization mechanism with the main objective of extending the lifetime of a WSN under restrictions related to energy consumption, information routing timeframes and resource usage.

5

Inheritance-based Adaptive Protocol for WSN Information Routing

The inheritance based-adaptive protocol for WSN information routing is one of the proposed routing solutions that uses a dynamic clustering mechanism of networks. Its implementation covers three main stages: first, a deployment and network initialization stage in which sensor nodes are deployed randomly in the operational environment. There is then a cluster construction stage, which is the heart of the mechanism and consists of structuring the network into a set of areas, i.e. clusters, in which there are master nodes (CHs) that guarantee coordination of all member nodes (MNs) and processing. CHs are able to communicate directly with the base station or over several hops by passing through other master nodes acting as relays. Finally, there is a communication stage that corresponds to the transmission of data according to pre-established routing protocols. The objective is thereby to select, by optimizing criteria, the best path for conveying data to the destination. The idea is to determine one or several performance functions that allow a range of parameters to be taken into account, such as the distances covered between the source and the destination, the time taken from beginning to end, the signal strength, the energy consumption and/or the number of hops.

5.1. Network deployment and initialization

Following a random or structured deployment of sensor nodes in an operational environment, the network initialization first goes through the establishment of an initialization message $Message\_Init\_SB$ in the manner of:

$$Message_{Init_{SB}}\left(\mathrm{ID}_{N_i},\mathrm{ID}_{N_j},P_{rob},N_{hops}\right) \qquad [5.1]$$

with ID_{N_i} and ID_{N_j} being the respective identifiers of source node N_i and destination node N_j, P_{rob} being a probability and N_{hops} being the distance between a node and the base station.

First, the bases station specifies the number of clusters that the network must have, before beginning to broadcast initialization messages $Message_{Init_{SB}}$ throughout the entire network by randomly generating a probability P_{rob}, described in [4.1], to choose the initial CHs.

As stated above, the number of CHs n_{CH} corresponds to 10% of the total number of nodes n that make up the network, and is calculated as follows:

$$n_{CH}=(10\%)n \qquad [5.2]$$

Diffusion consists of transmitting the same message to several targets at the same time by setting the values of the target addresses to zero. In this way, as shown in Figure 5.1, the base station (BS) broadcasts the message $(\mathrm{ID}_{SB},0,P_{rob},0)$ and, once it has been intercepted by a node, the latter identifies the source and the information received (P_{rob}), and carries out various processes, with part of the information to be sent to the base station later before it re-diffuses the message to its neighbor. At the same time as this, it builds its own routing table.

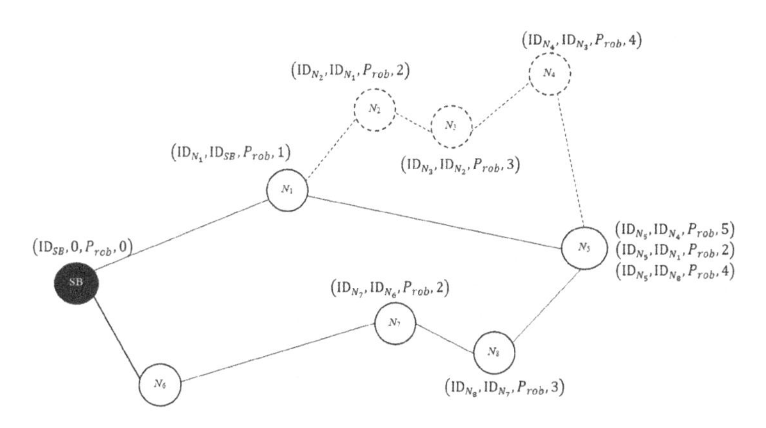

Figure 5.1. *Example of an initialization message being sent*

The processing first consists of increasing the number of hops (N_{hops}), such as:

$$N_{hops_{(i+1)}} = N_{hops_{(i)}} + 1 \qquad [5.3]$$

with i representing an index of a given sensor node.

This is a stage in which environment recognition is carried out on the immediate surroundings and the distance between the nodes and the base station.

NOTE 5.1.– Building a routing table is essential as it allows, where necessary, the best paths to be selected according to criteria specified beforehand for routing information to the destination (BS). We should note that the sensor node (N_5) detects several messages from different source nodes (N_1), (N_4) and (N_7). If the selected criteria concern the minimal number of hops, then (N_5) can reach the base station via a relay sensor node (N_1).

Moreover, a source node must have sufficient transmission power P_{T_x} to convey $Message\_Init\_SB$ to the target node. Depending on the distance d separating two nodes, P_{T_x} will influence the reception power P_{R_x} of the target node. In this way, on the basis of the Friis formula for transmitting messages in an open space [FRI 46], the strength of the received signal decreases in a quadratic manner in accordance with the distance of the transmitter:

$$P_{R_x} = P_{T_x} G_{T_x} G_{R_x} \left(\frac{\lambda}{4\pi d} \right)^2 \quad [5.4]$$

with G_{T_x} and G_{R_x} and λ respectively being the transmitter gain, the receiver gain and the length of the wave.

In embedded architectures, the strength of the received signal is converted to RSSI[1] which, in turn, is defined as the ratio of the power received and the reference power P_{ref}. In general, the reference power represents an absolute value of $\left(P_{ref} = 1\, mW \right)$.

The RSSI can also be formulated as follows:

$$RSSI = 10 \log \left(\frac{P_{R_x}}{P_{ref}} \right) [dBm] \quad [5.5]$$

As stated, the greater the reception power, the more the RSSI increases. On the other hand, the greater the distance d separating the source and target nodes, the more the RSSI decreases.

Another generic model for calculating RSSI called *log-normal shadowing* [RAP 96] has been developed and is described using:

$$RSSI(d) = P_{T_x} - PL(d_0) - 10\eta \log_{10} \frac{d}{d_0} + X_\sigma \; [dBm] \quad [5.6]$$

where P_{T_x} and $PL(d_0)$ respectively represent the transmission power and the signal loss for a reference distance d_0 in which the exponent is η. The

1 Received Signal Strength Indicator.

random variations are expressed in a Gaussian form with a zero mean and the variance $\sigma^2 : X_\sigma \sim \mathcal{N}\left(0,\sigma^2\right)$.

In TinyOS, for a CC2420 radio component of a Mica2 sensor, the RSSI values provided by the physical layer under the ZigBee protocol are encoded on a byte averaging eight time-periods of approximately 128 μs. The strength of the calibrated radio signal is calculated by the following formula:

$$P = RSSI_{val} + Offset_{RSSI} \quad [dBm] \qquad [5.7]$$

where $Offset_{RSSI}$ is a corrective value defined empirically and estimated at approximately $-45dBm$.

In the following subsection, we will look at the usefulness of calculating the strength of the RSSI signal received.

5.2. Network architecture clusterization

After a random or structured deployment of sensor nodes in the environment, the cluster construction stage takes place, which is central to the mechanism and consists of structuring the network into a set of areas in which CH master nodes guarantee the coordination of all MNs and processes. Each sensor node must generate a random number nb_a, which will be compared to the probability P_{rob} to determine whether it will be a CH or simply an MN:

$$A\,node(i)\,is : \begin{cases} Potentially\,CH\ if\ nb_a \leq P_{rob} \\ NM\ if\ not \end{cases} \qquad [5.8]$$

5.2.1. *Status diffusion*

Before broadcasting their new status to all neighboring nodes, all CH nodes first inform the BS by either communicating directly via a message $Message_{Adv_{CH}}$ in a single hop, or selecting optimal paths in the case of a multi-hop communication. For this purpose, through [5.7] and the corresponding routing tables, they determine which nodes have the minimal number of hops $N_{hops_{(i)}}$. The structure of the message is as follows:

$$Message_{Adv_{CH\left(ID_{CH_i},\ ID_{SB},\ header\right)}} \quad [5.9]$$

The ID_{CH_i} and ID_{SB} message components respectively represent the identifiers of the source node CH_i and of the BS.

The BS in turn recovers all messages and builds a temporary routing table that corresponds to all nodes, both CHs and ordinary nodes.

NOTE 5.2.– At this stage, all nodes, both CHs and ordinary nodes, can act as relays. Upon receiving a message, they transmit it directly to the destination in a single hop, by relaying the message in turn via other relay nodes.

Secondly, each CH node broadcasts to all neighboring nodes, i.e. CHs or MNs, via the diffusion of its new status in a small announcement message $Message_{Adv_{CH\left(ID_{CH_i},\ ID_{SB},\ header\right)}}$ containing its identifier ID_{CH_i} and a *header*. It uses a CSMA/MAC layer access method in order to avoid potential collisions and interference during exchanges. At the same time, the remaining sensor nodes are waiting for the message $Message_{Adv_{CH}}$.

Once the $Message_{Adv_{CH}}$ has been received, each non-CH sensor node determines its membership to a cluster by choosing the CH requiring the smallest amount of energy for transmission. The membership decision is based on the strength of the RSSI signal received from the $Message_{Adv_{CH}}$. The greater the intensity of the RSSI signal in relation to a source node, the less transmission energy it requires and therefore the more likely it is to be chosen to belong to its cluster. It must inform the chosen CH of its affiliation and become an MN by sending an affiliation packet such as $Message_{Join_{REQ}}$ by using the same MAC layer access method: CSMA (Figure 5.2).

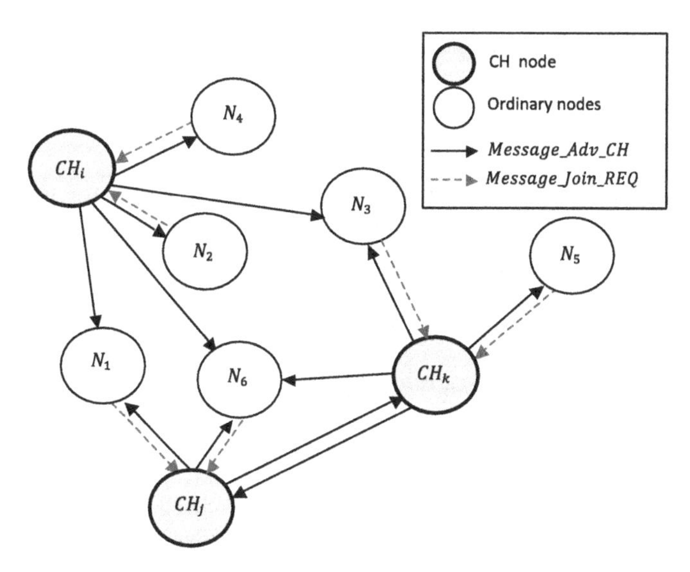

Figure 5.2. *Example of state and affiliation messages broadcast. For a color version of the figure, see www.iste.co.uk/touati/energymanagement.zip*

In contrast to LEACH, the information inheritance approach allows all nodes at the edge of the network to be reached, even if the signal strength of $Message_{Adv_{CH}}$ is reduced. In fact, multi-hop communication alleviates this problem and thus avoids the need for remote transmissions. The $Message_{Join_{REQ}}$ message is small and is structured as follows:

$$Message_{Join_{REQ(ID_{N_i},\, ID_{CH_j},\, header)}} \quad [5.10]$$

As shown in Figure 5.2, the sensor node (N_3) decided to send an affiliation message to the $(CH_k)node$ rather than the (CH_i) node as the strength of the RSSI signal indicator received from the (CH_k) is greater. This is confirmed by comparing the distances separating the (N_3) from the (CH_k) and (CH_i).

Outliers: If, for instance, a sensor node calculates the same RSSI value for two CHs, it must make a decision on its membership of a cluster. It may consult its routing table, which is initially formed during the recognition stage, to confirm the number of hops (N_{hops}) related to each of the two CHs required to reach the BS. The CH with a lower (N_{hops}) is selected by sending it an affiliation message as described in [5.10]. In the event that the number of hops is identical, it will randomly choose a cluster to which it belongs.

NOTE 5.3.– During status diffusion by CH nodes, none of them respond to others as they are already considered to be master nodes. They therefore auto-organize themselves in their routing tables as neighboring CHs. This would allow, if necessary, inter-CH multi-hop communication to transmit information to the BS. In our case, for instance, the sensor node (CH_k) can reach the BS by passing through the sensor node (CH_j).

Once the clusterization stage has been completed and all of the CH nodes have broadcast their statuses and received messages of affiliation, a stage of diffusing slots to all MNs at the level of each cluster is carried out by the corresponding CHs.

5.2.2. *Slot assignment*

The assignment of slots by CH nodes is carried out at the MAC layer by a task scheduling based on the TDMA method where every CH assigns each of its MNs a time slot during which the node can communicate its information to the master node (Figure 5.3).

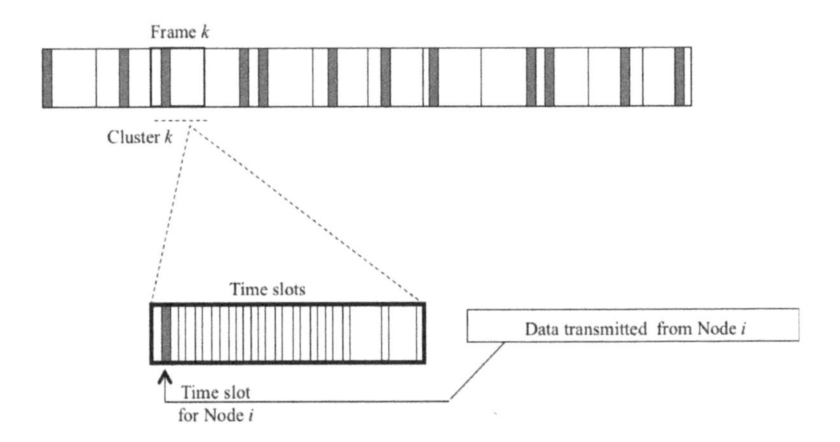

Figure 5.3. *Time slot assignment in a cluster. For a color version of the figure, see www.iste.co.uk/touati/energymanagement.zip*

The set of slots assigned to nodes belonging to a group constitutes a Frame, whose duration varies depending on the number of nodes in the cluster. This stage also establishes the distribution of CDMA codes to avoid interference and collisions between nodes in adjacent clusters. This is the case in the TRAMA[2] protocol that randomly prescribes two periods allowing the nodes to access the MAC layer randomly or in a scheduled manner [RAJ 03].

5.3. Data transmission and processing

This stage consists of collecting information, i.e. environmental information, through MNs, then processing it locally before conveying it during the allocated time to the related CH nodes (Figure 5.4).

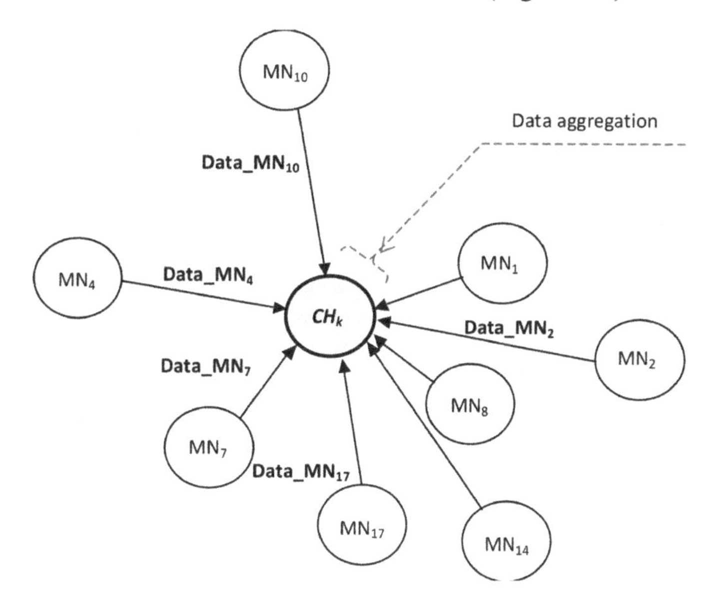

Figure 5.4. *Data aggregation in a cluster*

Once the information has been transmitted, each MN activates a standby task to optimize its energy. Within clusters, each CH carries out the fusion of information through aggregation of **Data_NM$_i$** data, collected from each of

2 TRaffic-Adaptive Medium Access.

its MNs by calculating the average before transmitting the result to the base station in one or multiple hops through route optimization.

NOTE 5.4.– It is important to highlight that sending **Data_NM$_i$** data using a set of MNs follows the scheduling principle based on the TDMA method.

5.3.1. *Optimization of CH-BS paths*

Optimizing paths consists of guaranteeing data conveyance between CH nodes and the base station on the basis of optimized cost functions. The aim is therefore to route the data by selecting the best path based on minimizing a performance function in accordance with one or several parameters to be optimized, such as the distances between source and target, the end-to-end timeframe, the signal strength, the energy consumption and/or the number of hops.

The lifetime of a WSN can be divided into a set of rounds or cycles in which each MN or CH sensor node operates over several stages (Figure 5.5(a) and (b)). During every round, each MN generates a data packet that it transmits to its CH while respecting the allocated time.

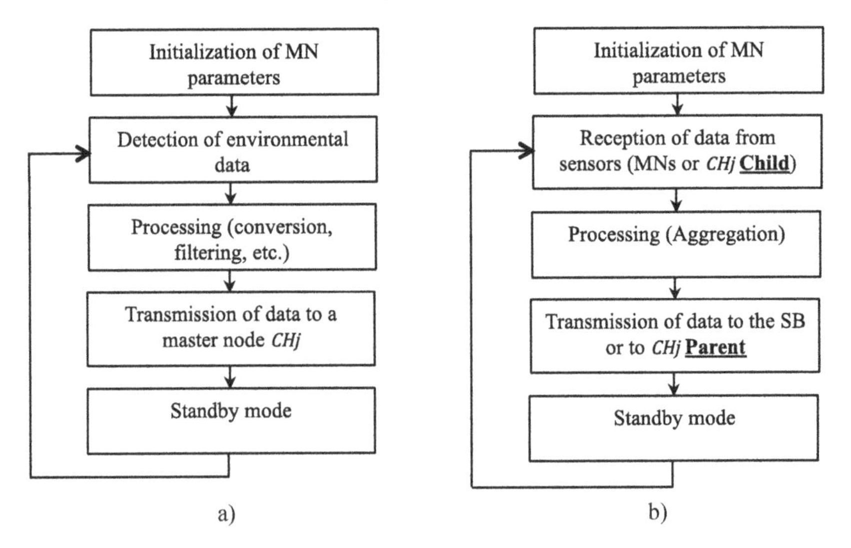

Figure 5.5. *Operating mode of an a) MN and a b) CH*

Each CH carries out information fusion by aggregating all data collected from MNs before transmitting it either directly to the BS according to the CSMA/CA access method of the MAC layer, or indirectly via other parent CH nodes. Once the transmission has occurred, the CHs put themselves in standby mode to wait for other data emerging from their own MNs or other child CHs.

Child–parent communication includes the selection of the best path to reach the BS. For this purpose, it is essential to define and then minimize a performance function by optimizing its parameters and observing their evolution over time. Among these vital parameters, we can note, for instance, the distance between source and target, end-to-end timeframe, the signal strength and the energy consumption. On the basis of the consumption model presented in Subsection 4.2.2, we can determine two kinds of model depending on whether the node is a CH or an MN.

In the case of a cluster, therefore, the set of MNs does not receive messages in the form of packets but, on the contrary, they transmit data to the corresponding CHs by consuming, in each round, energy in the form of:

$$E_{T_x(N)}\left(k,d_{ij}\right)=k_N E_{elec}+k_N E_{amp}\left(d_{ij}\right)^m \qquad [5.11]$$

with d_{ij} being the distance between the MN_i and the CH_j, and m being a path loss parameter. Depending on the environment, i.e. transmission of k_N packets in the open or in a constrained area, it can take on values of 2 or 4 respectively [RAJ 03].

At the level of every cluster, each CH transmits k_{CH} packets directly to the base station or in multiple hops to a CH_{PARENT} consuming $E_{T_x(CH)}$ energy, such as:

$$E_{T_x(CH)}\left(k,d_j\right)=k_{CH} E_{elec}+k_{CH} E_{amp}\left(d_j\right)^m \qquad [5.12]$$

The d_j parameter is the distance separating two consecutive CHs, or a CH and a BS.

Upon reception, the consumption at the level of the CHCHILD is around:

$$E_{R_x(CH)}\left(k_{R_{CH}},k_{R_N}\right)=\alpha k_{R_{CH}}E_{elec}+\left(1-\alpha\right)k_{R_N}E_{elec} \qquad [5.13]$$

where α is a variable taking the value of 0 or 1, depending on the transmission stages. In intra-cluster, the value of α is 0 and, in inter-cluster, it is 1.

To sum up, the energy consumed at the level of each CH during the reception of data packets can be described as follows:

$$E_{R_x(CH)}\left(k_{R_{CH}},k_{R_N}\right)=\begin{cases}k_{R_{CH}}E_{elec} & si\left(\alpha=1\right)\\ k_{R_N}E_{elec} & si\left(\alpha=0\right)\end{cases} \qquad [5.14]$$

In this way, in each round, the total energy consumed at the level of an MN (i) belonging to a given cluster (j) only concerns the transmission energy defined in [5.12]. This is written as follows:

$$E_N\left(i,j\right)=k_N E_{elec}+k_N E_{amp}\left(d_{ij}\right)^m \qquad [5.15]$$

Similarly, the total energy consumed in each round at the level of a CH (j) is around:

$$E_{CH}\left(j\right)=k_{CH}E_{elec}+k_{CH}E_{amp}\left(d_j\right)^m+\alpha k_{RCH}E_{elec}+\left(1-\alpha\right)k_{RN}E_{elec} \qquad [5.16]$$

Moreover, the processing of information by processors in WSNs requires an energy consumption E_{proc} close to 15% of the overall energy. This consumption is due mainly to energy loss during switching E_{commut} and current leakage E_{leak}. The energy consumed by a sensor node during each round for processing information of k_N length can therefore be written as follows:

$$E_{proc(N)}\left(k_N,N_{Cyc}\right)=k_N N_{Cyc}C_{avg}V_{sup}^2+k_N V_{sup}I_0 e^{\frac{V_{sup}}{n_p V_t}}\left(\frac{N_{Cyc}}{f}\right) \qquad [5.17]$$

Given that the processing and aggregation of information is carried out solely at the level of CH nodes, the energy used during each round can be formulated as:

$$E_{proc(CH)}\left(k_{CH}, k_N, N_{Cyc}\right) = k_{CH} E_{proc(N)}\left(k_N, N_{Cyc}\right) \quad [5.18]$$

where N_{Cyc} is the number of clock cycles, C_{avg} the average capacitance, I_0 the current leakage, n_p the constant depending on the type of processor used, V_t the thermal voltage, V_{sup} the supply voltage, f the sensor frequency and k_{CH} the length of information at the level of the CH node.

To determine the total energy consumed in each round by all MNs and CHs, we can write:

$$E_{tot} = n_{hops} \sum_{j=1}^{k} \left(E_{proc(CH)}\left(k_{CH}, k_N, N_{Cyc}\right) + E_{CH}(j) + \sum_{i=1}^{n} E_N(i,j) \right) \quad [5.19]$$

The parameters n and k respectively represent the number of MNs associated with a given cluster (j) and the number of clusters in the network. As shown in Figure 5.6, conveyance of the data aggregated at the level of a CH_{CHILD} to the BS is carried out directly in a single hop and, in this case, the CH_{PARENT} is represented by the BS or relayed in several hops for which each CH_{PARENT} receiving a message from the source CH_{CHILD} becomes a CH_{CHILD} in turn by routing the message to its destination.

In contrast to the consumption model described previously, which considers parent CHs and the BS separately, thereby complicating implementation of the model, we present a new energy model developed in the LIASD laboratory, whose principle is the decomposition of distances into a set of non-linear and non-identical segments.

The message sent by the source CH_{CHILD} to the target BS via several CH_{PARENT}, each in turn becoming a CH_{CHILD}, covers several segments of r_{ij} distances whose final distance can be expressed by:

$$d_{ij} = \frac{k-1}{n_{hops}} \left(\sum_{j=1}^{k-1} r_{ij} \right) \quad [5.20]$$

with n_{hops} being the number of hops required to transport a message from a CH_{CHILD} node to the BS and k being the total number of clusters in a given round.

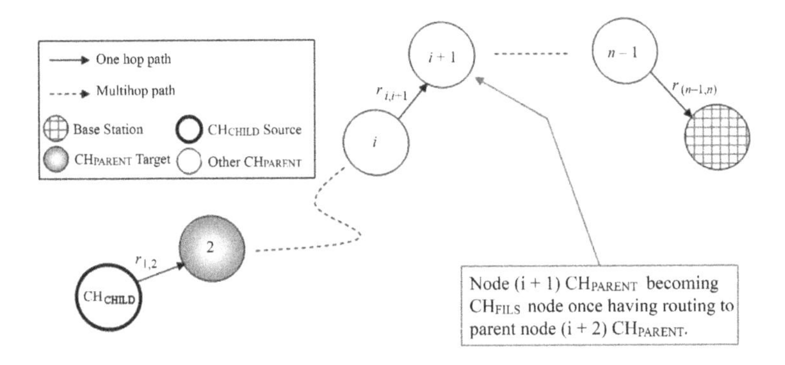

Figure 5.6. *Multi-hop CH_{CHILD} – CH_{PARENT} communication. For a color version of the figure, see www.iste.co.uk/touati/energymanagement.zip*

The optimal number of hops $\left(n_{hops\_opt}\right)$ can be calculated as follows:

$$\frac{dE_{tot}}{dn_{hops}} = 0 \qquad [5.21]$$

By replacing $(m = 2)$ in [5.21], the result is:

$$n_{hops_{opt}} = \frac{\left(E_{amp}\left(k-1\right)^2 \sum_{j=1}^{k-1}\left(r_{ij}\right)^2\right)}{\left(kk_p\left(E_{proc(N)}\left(k_N, N_{Cyc}\right)+\left(1+n\right)E_{elec}\right)+ E_{amp}\sum_{j=1}^{k}\sum_{i=1}^{n}\left(d_{ij}\right)^2\right)} \qquad [5.22]$$

The greater the number of clusters, the more $\left(n_{hops\_opt}\right)$ increases, and the CH_{CHILD} source node can therefore have more choice in routing messages to the target BS via the CH_{PARENT} nodes.

Moreover, if the amount of energy necessary for transmitting a message is around the same as the reception energy, particularly when the inter-CHs r_{ij} distances are short, the single-hop transmission is more desirable than multi-hop routing (Figure 5.7). It is clear that the shorter the distances are, the larger the number of hops becomes, thereby involving more CH_{PARENT} nodes for processing messages. The costs, in terms of energy consumption, become greater. If the target is covered, it is preferable for the CH_{CHILD} node to transmit directly to the target BS, instead of passing through several CH_{PARENT} nodes.

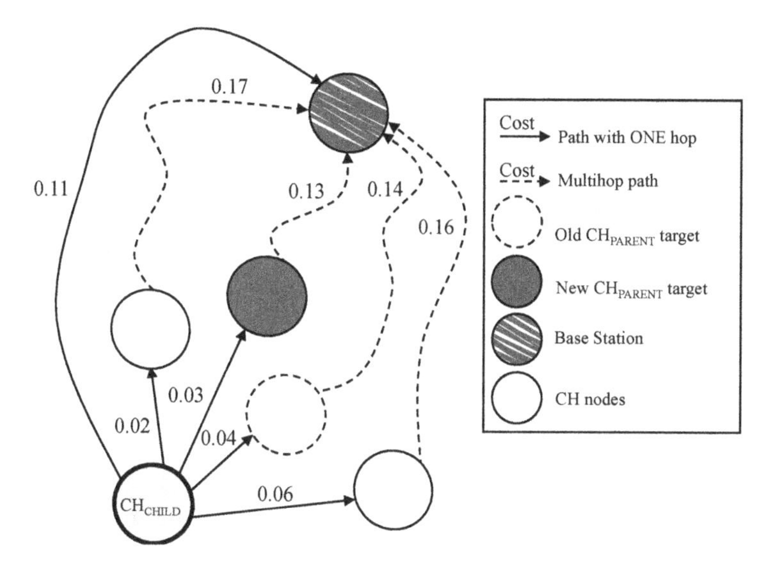

Figure 5.7. *Multi-hop CHCHILD – BS communication. For a color version of the figure, see www.iste.co.uk/touati/energymanagement.zip*

For a single hop, the necessary cost can reach 0.11 [UoV[3]], while for several hops the cost can vary between 0.13 and 0.17 [UoV]. If this is not the case, it is preferable to choose the best path for which the number of hops is optimal, i.e. $\left(n_{hops\_opt}\right)$ (Figure 5.8).

Thus, the CH_{CHILD} node determines the maximum value of the residual energy $E_r^{(j)}$ for all CH_{PARENT} nodes belonging to its surroundings, before calculating an energy threshold $E_{r-threshold}$, which should correspond to 20% of the maximum residual energy.

3 Unit of Value.

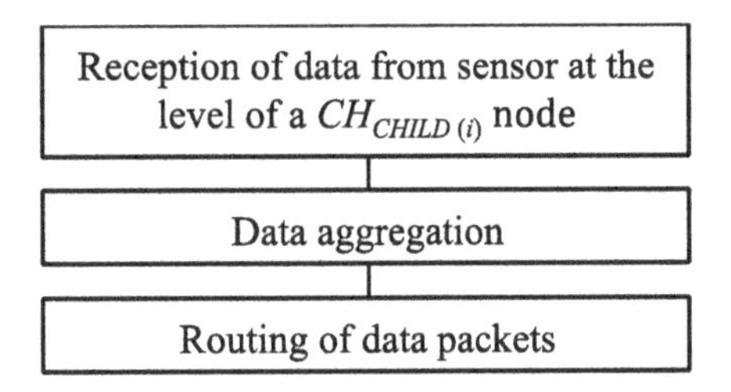

Figure 5.8. *Simple routine for selecting optimal paths*

In the case of a CH_{CHILD} node whose CH_{PARENT} node number is k' and assuming that CH_{PARENT} node h has the greatest residual energy, which can be calculated as follows:

$$E_r^{(h)} = max\left(E_r^{(1)}, E_r^{(2)}, \ldots, E_r^{(k')}\right) \qquad [5.23]$$

The energy threshold can therefore be determined as:

$$E_{r-seuil} = (20\%)\, E_r^{(h)} \qquad [5.24]$$

The best choice will focus on the CH_{PARENT} node h, for which the energy must respect the following condition:

$$\left(E_{r-PARENT}^{(h)} - E_{r-PARENT}^{(j)}\right) \geq E_{r-threshold} \qquad [5.25]$$

If this is not the case, the choice will center on the CH_{PARENT} node with a number of hops $\left(n_{hops}\right)$ equal or closer to the $\left(n_{hops\_opt}\right)$ determined by equation [5.22]. On the assumption that the smallest number of hops to reach the BS lies with the CH_{PARENT} p node, we use:

$$dn_{hops}^{(p)} = min\left(\left|n_{hops}^{(j)} - n_{hops\_opt}\right|\right) \qquad [5.26]$$

However, if there are several redundant values $\left(dn_{hops}^{(p)}\right)$, the CH_{CHILD} source node bases its selection criteria on the maximum power received by each of the likely CH_{PARENT} target nodes by calculating the $RSSI_{PARENT}^{(j)}$ values, such as:

$$RSSI_{PARENT}^{(l)} = max\left(RSSI_{PARENT}^{(1)}, \ldots, RSSI_{PARENT}^{(k')}\right) \qquad [5.27]$$

The greater the RSSI value provided by each of the CH_{PARENT} target nodes, the closer it is to the source CH_{CHILD}. Less energy will therefore be used, as the transmission uses on average over 70% of the energy of a sensor node.

In the event that there are redundant RSSI values, the source CH_{CHILD} will proceed through a random choice to its target CH_{PARENT}.

5.3.2. *CH node selection*

Every time that a round, r, begins, the network communication architecture changes in accordance with the total energy of the network by taking into account the number of nodes remaining active and the number of CH nodes formed. In the child–parent information inheritance approach, with the exception of the initial stage of network construction and clusterization, both the formation of new clusters and the selection of corresponding CH nodes follow an adaptive mechanism shown earlier in which several parameters must be considered, namely the residual energy and the received signal strength as identified by the RSSI indicator.

There are two specific cases in this instance. The first concerns intra-cluster functioning where other MNs can take over their corresponding CH, and the second quite simply relates to a complete reorganization of the network. The idea is to minimize communication with the BS, as in the case of LEACH or all of its variants. This in fact allows the number of transmissions to be reduced and, consequently, the energy consumption to be optimized, thereby extending the lifetime of the network.

In intra-cluster operations, in each j cluster, and at the level of each round r, the selection of a new CH is carried out by the CH of the preceding round $(r-1)$. In other words, for any given cluster, a node member k, (NM_k) wishing to become CH for the next round r must have the largest amount of residual energy. This is shown in the following equation:

$$A\,node(k)\,becomes:\begin{cases} CH_k^{(j,r)}\ if\ E_{r_{NM_k}}^{(j,r-1)} > E_{r_{CH_i}}^{(j,r-1)} \\ NM_k^{(j,r)}\ if\ not \end{cases} \qquad [5.28]$$

Generally, an MN $k, \left(NM_k^{(j,r-1)} \right)$ becomes CH if its residual energy is greater than that of its previous CH $\left(CH_i^{(j,r-1)} \right)$, and if not it maintains its status as MN.

Once the data has been received at the level of the (CH_i) node, the latter determines the residual energy of all of its (NM_i) with $\left(i \in \left[1, n_j \right] \right)$ and n_j being the number of MNs belonging to the cluster j, before comparing its own energy value with that of its MNs in the following manner:

$$diff_{E_r}^{(j,r-1)} (CH_i, NM_k) = E_{r_{CH_i}}^{(j,r-1)} - E_{r_{NM_k}}^{(j,r-1)} \quad [5.29]$$

If $\left(diff_{E_r}^{(j,r-1)} (CH_i, NM_k) > 0 \right)$, the role of the (CH_i) node will be renewed for the round following r, and, if not, the $\left(CH_i^{(j,r-1)} \right)$ node determines the maximum value of the residual energies of each of its MNs:

$$E_{r_{NM_k}}^{(j,r-1)} = Max\left(E_{r_{NM_1}}^{(j,r-1)}, E_{r_{NM_2}}^{(j,r-1)}, \ldots, E_{r_{NM_i}}^{(j,r-1)}, \ldots, E_{r_{NM_{n_j}}}^{(j,r-1)} \right) \quad [5.30]$$

If there is no duplication, it attributes a new Cluster-Head status to the $\left(NM_k^{(j,r-1)} \right)$ node for next round. If there is duplication, the Cluster-Head from the preceding round $\left(CH_i^{(j,r-1)} \right)$ determines the likely future CH by calculating the RSSI indicators of all MNs with the same maximum value using the strength of the received signals, before selecting the member node with the biggest RSSI value.

For this purpose, we will consider k and k', two MNs belonging to the same cluster j with maximum residual energy, such as:

$$E_{r_{NM_k}}^{(j,r-1)} = E_{r_{NM_{k'}}}^{(j,r-1)} \quad [5.31]$$

Choosing which of k and k' is the best candidate to become CH requires the calculation and comparison of the corresponding RSSI values:

$$diff_{RSSI}^{(j,r-1)}\left(NM_k, NM_{k'}\right) = \left(RSSI_{NM_k}^{(j,r-1)} - RSSI_{NM_{k'}}^{(j,r-1)}\right) \qquad [5.32]$$

Therefore:

$$\begin{cases} NM_k \equiv CH\ if\ diff_{RSSI}^{(j,r-1)}\left(NM_k, NM_{k'}\right) > 0 \\ NM_{k'} \equiv CH\ if\ diff_{RSSI}^{(j,r-1)}\left(NM_k, NM_{k'}\right) < 0 \end{cases} \qquad [5.33]$$

NOTE 5.5.– If there are identical RSSI values, i.e. $diff_{\mathrm{RSSI}}^{(\mathrm{j,r-1})}\left(\mathrm{NM_k}, \mathrm{NM_{k'}}\right) = 0$, a new CH node will be chosen randomly. This is a very rare occurrence.

Every sensor node, be it a MN or a CH, must have enough energy to ensure that a contribution, even a minimal one, is made in the network. Thus, it must have a minimum amount of residual energy close to an energy threshold E_{T_x} that can be calculated in accordance with [5.24].

In fact, by having the E_{elec} and $E_{Tx_{amp}}$ parameters respectively representing the electronic transmission/reception energy and the amplification energy, the amplification factor ε_{amp} and the size of the packets to be transmitted, the threshold value can be determined in relation to each neighboring node. In each round r, the following comparison can be made:

$$E_{r_{node_i}}^{(j,r)} < E_{T_x} \qquad [5.34]$$

where $E_{r_{node_i}}^{(j,r)}$ and E_{T_x} respectively are the residual energy of a sensor node, i, during a round, r, and the energy threshold.

In this way, we can reformulate as follows:

$$a\,node(i)\,is: \begin{cases} excluded\ from\ the\ network\ IF\left(E_{r_{node_i}}^{(j,r)} - E_{T_x}\right) < 0 \\ renewed\ in\ the\ network\ IF\ not \end{cases} \quad [5.35]$$

The solution using child–parent inheritance must also take into account the overall functioning of the network and therefore of all nodes. The basic idea is to optimize not only the energy consumption in a WSN to maximize its lifetime, but also to minimize the number of messages exchanged between the different nodes, i.e. CHs and the base station. However, in the majority of routing approaches, the base station takes action at the end of each round to readjust the network, involving several message exchanges and therefore additional energy consumption caused by node-BS transmission.

Once the number of clusters j has been determined, the mechanism chooses the cluster that has the smallest number of MNs. It calculates a threshold value:

$$h^j = Min\left(N_{NM}^{(j)}\right) \quad [5.36]$$

with $N_{NM}^{(j)}$ being the number of member nodes belonging to a cluster j.

When all of the MNs in the cluster j have been chosen as CH nodes, the last CH warns the BS that the cluster no longer has the option of selecting a new coordinator node. The BS then begins to reorganize the network architecture by generating a new probability P_{rob} and the mechanism reruns itself by structuring the network once again.

5.4. Critical analysis and conclusion

In this chapter, we have presented a hierarchical routing solution based on information inheritance using an adaptive clusterization mechanism that allows significant improvements to be made in minimizing energy consumption, optimizing resources and therefore extending the lifetime of the network. This approach is based on three main stages: a stage of network deployment, initialization and recognition in which the sensor nodes with well-defined features specific to the type of application are deployed randomly in the operational environment, an adaptive clusterization stage in which the different states of the sensor nodes are taken into consideration, and a communication stage in which a single-hop or multi-hop child–parent communication concept is proposed to optimize the paths and the end-to-end timeframes for source–destination packet conveyance.

In contrast to the standard LEACH protocol, where the nodes choose their affiliated clusters several times, in the inheritance approach the node chooses the CH with the maximum amount of energy for sending a single message. The procedure for selecting new coordinator nodes does not pass by the base station, but is carried out directly at the level of each cluster. The former CH selects the new CH at the end of each round in accordance with its energy level initially, and, if not, according to the received signal strength when necessary.

Inheritance uses several metrics: the energy consumed, the network payload, the end-to-end timeframes for conveying packets and the strength of the signal from the RSSI indicator. The metric based on calculating RSSI values is used not only in the recognition and clusterization stages, but also in constructing optimal paths according the child–parent concept. The advantage is that the RSSI technology is widespread in the majority of WSN platforms, while also using relatively low-cost energy.

However, in WSNs, communication links are asymmetrical and, given that the strength of radio signals is measured upon reception, this can worsen the performance of the proposed routing mechanism and cause errors in estimating RSSI values. Those values can only be used to locate nodes after a certain period of time as the stability of the signal is only belatedly guaranteed. The time allocated for the recognition stage therefore becomes important, as it accelerates energy consumption and thereby minimizes the lifetime of the network.

6

Hierarchical Hybrid Routing: the HRP-DCM Solution

In this chapter, we will present an alternative based on the hybrid HRP-DCM[1] protocol, which allows improvements to be made at all levels, from the network recognition stage to the path optimization stage, during the exchange of information.

6.1. Introduction

Compared to TEEN or LEACH, implementing the HHRP routing approach based on dynamic clusterization mechanism significantly improves the performance of WSNs in minimizing energy consumption and optimizing resources, and therefore, also in prolonging the network lifetime. However, the use of RSSI measures at all levels of the mechanism, i.e. in the recognition of surroundings, clusterization and construction of optimal paths according to the child–parent concept, can cause the network to deteriorate due to the asymmetry of communication links. Moreover, the information can only be used at a certain time due to the stability of the RSSI signal. The time allocated for the environmental recognition stage therefore becomes important, as it accelerates energy consumption and thereby reduces the lifetime of the network.

1 Hybrid Routing Protocol based on the Dynamic Clustering Method.

Several methods that enable the quality of positioning to be improved have been developed and implemented for rolling out WSN applications [MUN 09]. These methods use AOA[2] measurements to locate an object in its operational environment, or other measures such as TOA[3] and TDOA[4], which are based on the flight time of information between a source and destination. The latter two are able to calculate distances very precisely, given the size of the bandwidth.

In this study, we propose a solution based on the HRP-DCM protocol, which uses the concept of calculating temporal distances during the environmental recognition stage. The aim is to shorten the delays allotted for the network initialization stage and therefore reduce energy consumption.

6.2. HRP-DCM routing mechanism

The HRP-DCM mechanism is an improved version of the child–parent information inheritance-based routing that circumvents problems caused by the asymmetry of communication links, among other issues. Its implementation has four significant stages to guarantee an exchange of information on the basis of optimization, namely energy consumption, resource use and timeframes for conveying information between source and target. We can draw attention to environmental recognition, clusterization, slot distribution and, lastly, communication, where optimal routes are constructed according to the child–parent concept [AOU 13a, AOU 13b, AOU 14a, AOU 14b, AOU 15a].

6.2.1. *Environmental recognition*

Once the nodes are randomly deployed in the operational environment, neighbors are discovered by exchanging messages between different sensor nodes. Each node $N_i \in H_s$ (with H_s being the starting set) broadcasts a

2 Angle Of Arrival.
3 Time Of Arrival.
4 Time Difference Of Arrival.

message that includes its identifier ID for all neighboring nodes L_s (the arriving set) in order for the latter to be able to recognize their surroundings. The message is:

$$Message_{Init_{SB}}\left(\mathrm{ID}_{N_i},\mathrm{ID}_{N_j},P_{rob},N_{hops},\Delta T_{H_s}^{(N_i)}\right) \quad [6.1]$$

with $\Delta T_{H_s}^{(N_i)}$ being the time variation during the transmission of a $Message_{Init_{SB}}$, such that:

$$\Delta T_{H_s}^{(N_i)} = T_{H_s}^{(N_i)} - T_0 \quad [6.2]$$

with T_0 and $T_{H_s}^{(N_i)}$ respectively being the initial *TIMER* value and the corresponding time when $Message_{Init_{SB}}$ is transmitted by the N_i node.

Each sensor from the arriving set N_j can therefore build a routing table by authenticating its closest neighbors with their respective identifiers ID_{N_i} and their depth in relation to the BS N_{hops} increased by a factor of 1 through the implementation of a mechanism for calculating the flight time between source and target (Figure 6.1). Similarly, a sensor node N_j belonging to the arriving L_s set receiving the same message should first calculate the temporal variation:

$$\Delta T_{L_s}^{(N_j)} = T_{L_s}^{(N_j)} - T_0 \quad [6.3]$$

with $\Delta T_{L_s}^{(N_j)}$ and $T_{L_s}^{(N_j)}$ being respectively the temporal variation in the reception of the $Message_{Init_{SB}}$ and the corresponding time when it is received by the ordinary N_j node.

The sensor node N_j then in turn broadcasts the $Message_{Init_{SB}}$ message to all neighboring nodes, even those that already exist in its routing table.

In contrast to the inheritance approach that uses the strength of radio signals to estimate the distance between sensor nodes, and over which asymmetrical communication links exist and are therefore able to negatively impact the performance of the routing mechanism, the HRP-DCM approach can circumvent this issue. Moreover, the RSSI data can only be used if it is stable, increasing the time allotted to the processing and calculation stage. We are able to determine the difference using [6.2] and [6.3]:

$$diff_{ij} = \Delta T_{L_s}^{(N_j)} - \Delta T_{H_s}^{(N_i)} = T_{H_s}^{(N_j)} - T_{L_s}^{(N_i)} \quad [6.4]$$

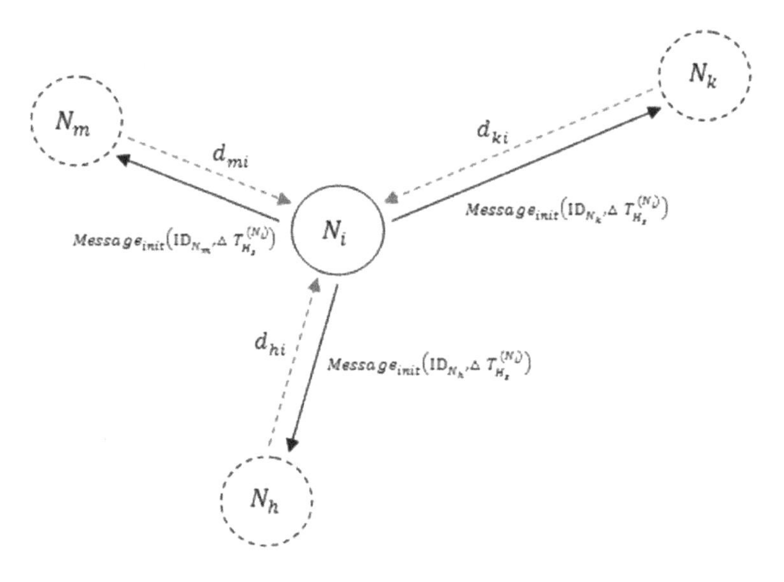

Figure 6.1. *Distance calculation mechanism*

Knowing the radio transmission frequency and the length of the messages transmitted and using the $diff_{ij}$ parameter representing the time it takes to transmit a message from the N_i node to the N_j node, we are able to locate the neighboring nodes. Once the recognition stage has been completed, all sensor nodes have a perfect knowledge of the immediate surroundings, namely the distances between different nodes d_{ij} and the depth N_{hops} in relation to the BS.

6.2.2. *Clusterization stage*

As in the inheritance-based approach, a sensor node can only become the coordinator if the random number nb_a that it generates is less than or equal to the probability P_{rob} fixed by the BS. Once the BS has been informed of its new status through a $Message\_Adv_{CH\left(ID_{CH_i}, ID_{SB}, header\right)}$ message, directly in a single hop if the BS is directly reachable or in several hops, in this instance, the choice will focus on the nodes with a minimal $N_{hops_{(i)}}$ depth, all coordinator nodes again broadcast their status by sending a warning message $Message\_Adv_{CH\left(ID_{CH_i}, header\right)}$ containing its identifier ID_{CH_i} and a *header* to specify that it is an announcement message.

Once the $Message\_Adv_{CH}$ message has been received, each non-CH sensor node determines its affiliation to a cluster by selecting the CH that requires the least transmission energy. The affiliation decision is based on a minimal distance which separates the different coordinator nodes and the sensor nodes (Figure 6.2).

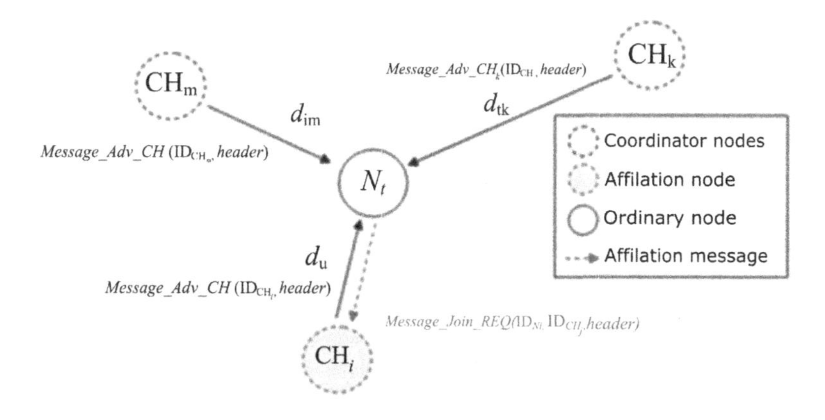

Figure 6.2. *Node affiliation to a cluster*

Upon reception, each sensor node chooses which coordinator node to belong to by sending an affiliation packet such as $Message_{Join_{REQ(ID_{N_i}, IDCH_j, header)}}$ in accordance with the CSMA MAC access layer.

Once the $Message\_Adv_{CH(\text{ID}_{CH_i}, header)}$ messages from different coordinators have been received at the sensor node level, the sensor node will check its routing table and choose a cluster with a minimal time distance to which to belong. If, for instance, there are redundant values, the choice will focus on the number of hops (N_{hops}). The CH that requires the smallest number of hops to reach the BS will be chosen. As in the inheritance-based approach, if the number of hops is identical, the choice of cluster affiliation will be made at random.

6.2.3. *Slot distribution*

In order to endure organized intra-cluster communication and avoid interference and collisions between nodes, each coordinator node provides its different NMs with well-defined time slots. This stage is similar to the stage described in the previous chapter.

6.2.4. *Communication stage*

Transmitting information from a source node to the BS first passes by the coordinator node, following data processing and aggregation, and the corresponding CH_{CHILD} must choose the best path by optimizing constraints such as the distances covered between source and target, the end-to-end delays, the signal strength, the energy consumption and/or the number of hops.

The pseudo-code (Figure 6.3) related to selecting the optimal route uses the same consumption model presented in subsection 4.2.2 and is as follows:

```
FOR EACH cluster j(j∈[1:k']) belonging to the surroundings of the CH_CHILD source node
    Calculate the maximal residual energy E_{r-PARENT}^(h) such as h∈[1:k']
        Determine the energy threshold E_{r-seuil}
        Test the energy in relation to the threshold
        IF (E_{r-PARENT}^(h) - E_{r-PARENT}^(j)) ≥ E_{r-threshold}
                Select the h node as CH_PARENT target node
        OTHERWISE
                Determine the number of hops (n_hops^(j)) of each CH_PARENT
                target node
                Test the difference between (n_hops^(j)) and (n_hops_opt)
                IF there is only one dn_hops^(p) solution
                        Select the p node as CH_PARENT target node
                OTHERWISE
                        There are several redundant solutions
                Compare the temporal distances of CH_PARENT target
                nodes diff_CH_j (j=1:k)
                        Calculate the minimum value:
                        diff_CH_l = min(diff_CH_1, diff_CH_2, ..., diff_CH_j, ..)
                        IF diff_PARENT^(l) e xists
                                Select the l node as CH_PARENT target node
                        OTHERWISE
                                There are several redundant solutions
                                Random choice of CH_PARENT target node
                        END
                END
        END
END
```

Figure 6.3. *Pseudo-code for selecting the optimal path (HRP_DCM)*

However, if there are several redundant values $\left(dn_{hops}{}^{(p)}\right)$, the CH$_{CHILD}$ source node bases its selection criteria on the minimum temporal distance $diff_{CHILD-PARENT}{}^{(j)}$, such that:

$$diff_{CHILD,PARENT}{}^{(l)} = min\left(diff_{CHILD,PARENT}{}^{(1)}, .., diff_{CHILD,PARENT}{}^{(k')}\right) \quad [6.5]$$

with:

$$diff_{CHILD,PARENT}{}^{(k')} = T_{H_s}^{(CHILD)} - T_{L_s}^{(PARENT)} \quad [6.6]$$

The smaller the temporal distance provided by each of the CH$_{PARENT}$ target nodes, the closer it is to the source CH$_{CHILD}$. The amount of energy spent will therefore be lower as the transmission uses on average 70% more energy than a sensor node. In this routing approach, fewer calculations are carried out to select the target CH$_{PARENT}$ and the decision can be taken quickly, as each CH$_{CHILD}$ already has a database of its entire surroundings that was established during the initial reconnaissance stage.

Each sensor node i belonging to a cluster j must have enough energy to transmit or process information in the network. In each round r, it must have a minimum residual energy $E_{r_{node_i}}^{(j,r)}$ close to an energy threshold E_{T_x}, such that:

$$E_{r_{node_i}}^{(j,r)} < E_{T_x} \quad [6.7]$$

Therefore, as in HHRP:

$$a\,node(i)\,is: \begin{cases} excluded\ from\ the\ network\ IF\left(E_{r_{node_i}}^{(j,r)} - E_{T_x}\right) < 0 \\ renewed\ in\ the\ network\ IF\ not \end{cases} \quad [6.8]$$

The HRP-DCM approach determines which cluster has the smallest number of NMs in the following way:

$$h^j = Min\left(N_{NM}^{(j)}\right) \quad [6.9]$$

with $N_{NM}^{(j)}$ being the number of member nodes belonging to cluster j.

This allows for the estimation of the smallest number of rounds required to ensure the operation of the network before the reorganization stage. The last CH ensuring the coordination alerts the BS that the cluster cannot select a new coordinator node. The BS therefore begins to reorganize the network architecture by generating a new probability P_{rob} and the mechanism restarts itself by clusterizing the architecture again.

6.3. Conclusion

We have just presented an HRP-DCM protocol solution, which is an alternative to the inheritance-based solution, and allows improvements to be made in terms of energy, resource use and processing time. Its implementation can avoid the use of RSSI measures at all levels of the mechanism, i.e. environmental recognition, clusterization and construction of optimal paths according to the child–parent concept. The RSSI data can in fact be the source of deterioration in the network due to the asymmetry of communication links, requiring the data to be used slowly as the RSSI signal takes a long time to stabilize.

The HRP-DCM solution is based on calculating temporal distances during network deployment and environmental recognition. This idea enables the times allocated to the initialization stage to be shortened, thereby reducing energy consumption.

The next chapter will provide a comparative study of different routing protocols dedicated to WSNs with a hierarchical architecture, such as HRP-DCM, the inheritance-based routing known as HHRP[5], TEEN and LEACH.

5 Hybrid Hierarchical Routing Protocol.

7

Performance Evaluation

In this section, we are going to carry out a comparative study of different routing protocols dedicated to WSNs with a hierarchical architecture, such as HRP-DCM, the inheritance-based routing known as HHRP, TEEN and LEACH. The metrics used concern energy consumption, end-to-end conveyance delays and the flow of information exchanged.

7.1. Introduction

This book addresses the issue of deploying and implementing a WSN by focusing on the methods and mechanisms that improve efficiency in the functioning of the network. The aim is to provide significant improvements in terms of energy consumption and optimizing resource usage, thereby extending the lifetime of the network. Using a material solution alone is not enough to guarantee WSN performance. It is therefore necessary to turn to software solutions, which will allow information use to be controlled from its source through its conveyance to its destination by taking into account the inherent features of sensors and the related energy restrictions. Partially responding to these needs involves the development of computing tools and protocol strategies in low-consumption mode, and the implementation of mechanisms based on information routing techniques. For this purpose, we have presented the latest developments in the situation and noted two protocol solutions intended for WSNs routing information with a hierarchical architecture: HHRP and HRP-DCM, which we will examine alongside LEACH and TEEN solutions on the basis of a number of metrics, such as energy consumption, end-to-end conveying delays and the flow of information exchanged.

7.2. Experimental platform

The evaluation platform is a dedicated open source for embedded applications such as TinyOS[1] version 2.x installed on a GNU/LINUX OS with a set of packages to be used [LEV 03, LEV 04, LEV 09].

The development part is based on a programming language focused on components, syntactically similar to C language, *NesC* [GAY 03], which offers the possibility of drastically reducing the size of the system's memory and applications, i.e. components. Using it requires three modules: interfaces, the configuration model and components.

The sensor model is a Mica2 from Crossbow (Figure 7.1), which has a low-power ATmega128L microcontroller and a CC2420 radio module. It also has two flash drives of 512 and 128 Kbytes respectively for storing the measurements and the program. The radio communication uses the ZigBee protocol, which can reach a range extending from 30 m in an enclosed environment to around 100 m in an outside environment.

– 2.4 GHz for enabling low-power WSN
– IEEE 802.11.4 compliant Radio frequency transceiver
– 51-pin expansion connector
– High speed (250 Kbps)
– Hardware security (AES-128)

Figure 7.1. *Mica2 sensor with specifications*

The MTS400CA sensor board has a set of sensors able to collect information on acceleration, atmospheric pressure, ambient light, temperature and humidity.

7.3. Choice of initialization parameters

Various experiments have been carried out by taking into account the inherent parameters of sensors, and we have considered several networks with different densities (Table 7.1).

1 Tiny-Operating System.

Parameters [units]	Values
Environmental surface $\left[m^2\right]$	100×100
Densities [n]	20, 50, 100, 150, 200, 500
Position of the BS $\left[m,m\right]$	(0.0)
Radio range $\left[m\right]$	25
E_{elec} $\left[nJ / bit\right]$	50
$d_{crossover}\left[m\right]$	50
ε_{fs} $\left[\frac{nJ}{bit} / m^2\right]$	1.29×10^{-15}
ε_{tr} $\left[\frac{nJ}{bit} / m^4\right]$	0.1×10^{-10}
k_{CH} $\left[bytes\right]$	29
k_N $\left[bytes\right]$	29
$k_{R_{CH}}$ $\left[bytes\right]$	29
k_{R_N} $\left[bytes\right]$	29
Loss of path [m]	2.4
Initial energy $\left[Joules\right]$	20
n_{CH}	$(10\%)n$
$Offset_{RSSI}$ $\left[dBm\right]$	-45

Table 7.1. *Experimental parameters*

7.4. Implementation and analysis of results

Performance evaluation includes the consideration of optimality criteria related to transmission and information processing.

The energy consumption graphs in Figure 7.2 perfectly illustrate the contribution of the two proposed protocol solutions in comparison to the LEACH and TEEN protocols. The HRP-DCM protocol solution in fact guarantees a better energy optimization than the HHRP, TEEN and LEACH approaches. This demonstrates that the improvements provided during the network initialization stage and, in particular, the neighborhood recognition in HRP-DCM enable the optimization of processing and radio transmissions for creating clusters, thereby limiting the number of messages exchanged between the different sensor nodes.

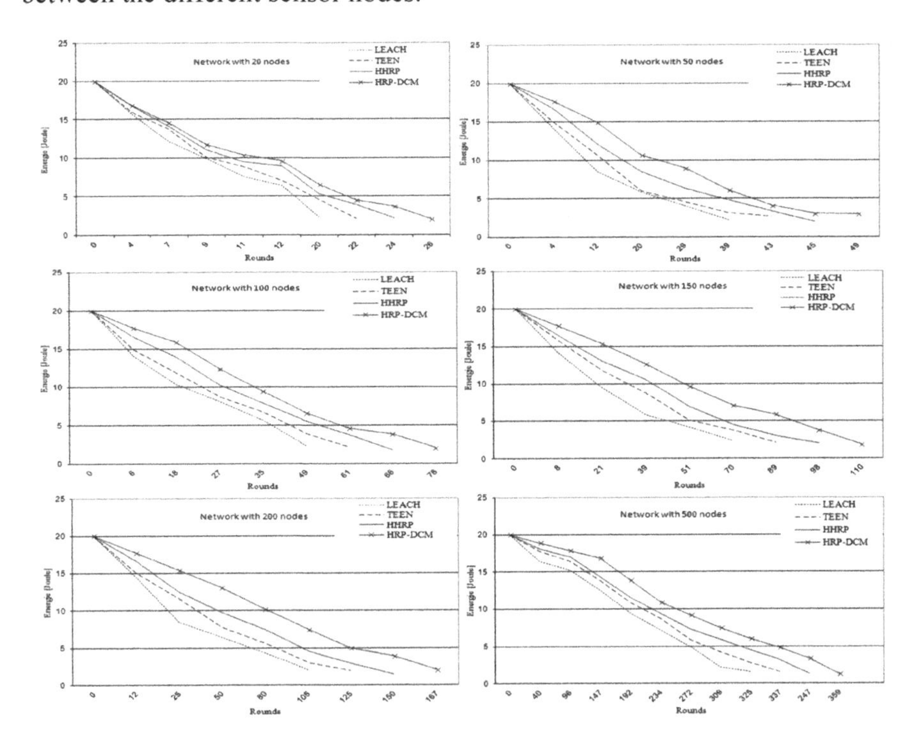

Figure 7.2. *Energy consumption*

If we consider that the performance of a WSN cannot be guaranteed after it loses 50% of its total energy, then, in the example of a network with 150 nodes and an average energy of 20 J, the HRP-DCM and HHRP solutions extend the network's operational lifetime to 53 and 42 rounds respectively, while it is only extended to 21 and 34 rounds respectively for TEEN and LEACH. This difference increases over time to reach the following values: 116, 100, 92 and 72 respectively for HRP-DCM, HHRP, TEEN and LEACH.

This shows that optimizing the initialization stage in terms of the number of transmissions, processing and allocated time allows the HRP-DCM protocol solution to improve the performance in energy consumption and resource use.

Moreover, integration of the nonlinear model, which allows the distances separating the source and target nodes to be determined according to the number of hops, with the proposed energy model significantly improves the optimization of energy consumption. The best compromise between the distances and the number of hops must correspond to the optimal path that has to convey information.

As shown in Figure 7.3, the conveying of aggregated data from a CH_{CHILD} to a BS is carried out directly in a single hop, and, in this case, the CH_{PARENT} is represented by the BS or relayed in several hops for which each CH_{PARENT} receiving a message from the source CH_{CHILD} itself becomes a CH_{CHILD} again by routing the message to its destination in turn.

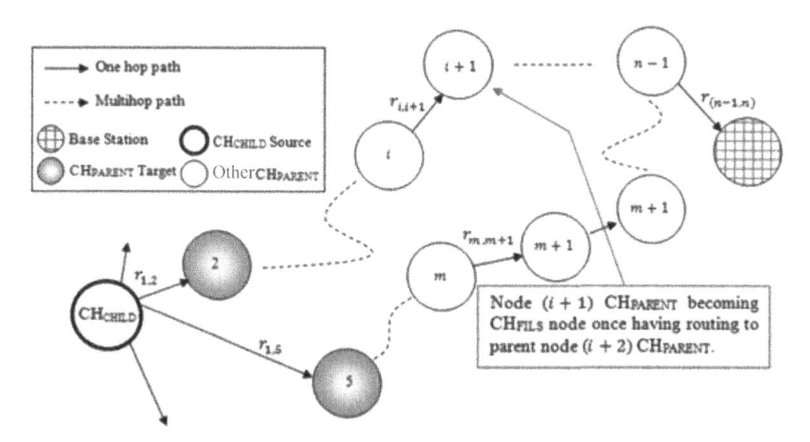

Figure 7.3. *Optimization of CH_{CHILD}–CH_{PARENT} communication. For a color version of the figure, see www.iste.co.uk/touati/energymanagement.zip*

The idea is to improve the longevity of the network by determining the best path that brings together the number of hops and the best-suited sensor nodes with the smallest distances. This network longevity can be illustrated by the various graphs in Figure 7.4. We should note that for a WSN with 50 nodes, for example, the network starts to lose its first nodes at the beginning of the first and second rounds respectively during implementation of the LEACH and TEEN protocols. As for the two proposed HHRP and HRP-DCM solutions, the first nodes disappear after 5 rounds. Moreover, the density is reduced over time until the network is lost completely, but this occurs more slowly with the HRP-DCM solution.

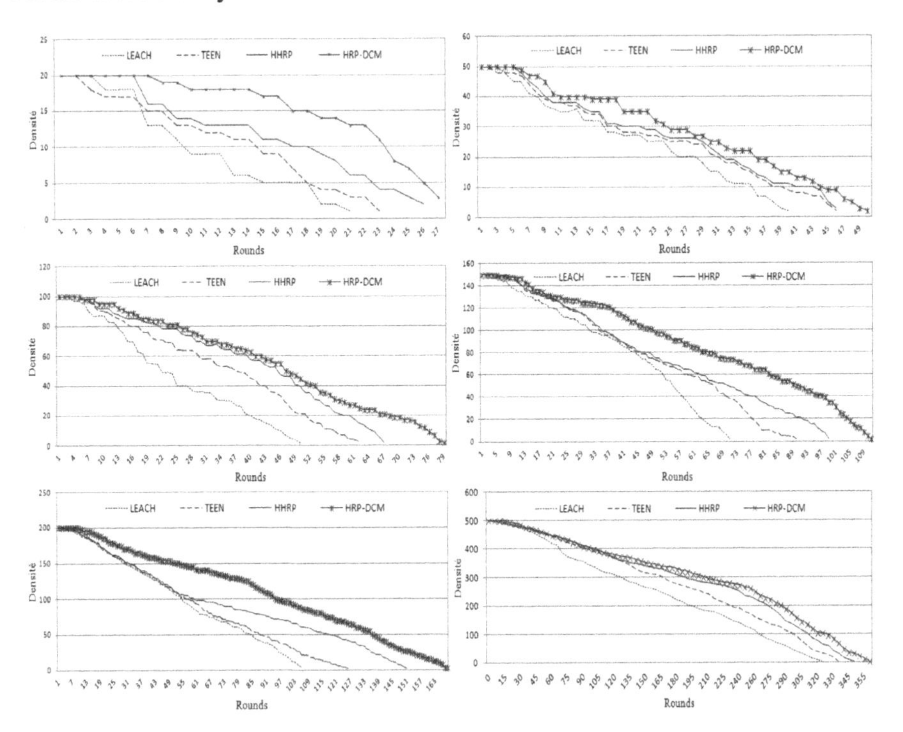

Figure 7.4. *Network longevity*

In this way, if we always consider that WSN performance cannot be guaranteed after the loss of 50% of the nodes that form it, then we can conclude that HHRP and HRP-DCM improve the network performance, by extending the longevity of the network to 28 and 31 rounds respectively, in contrast to LEACH and TEEN which only reach 23 and 26 rounds respectively.

Other results that have been obtained confirm this (Table 7.2). In fact, the total energy consumed during the implementation of different protocols over precise execution times ranging from 300 to 3,600 seconds show that the HRP-DCM solution optimizes network operations by an average gain factor equivalent to 10%.

Execution time [s]	Total energy consumed [J]			
	LEACH	TEEN	HHRP	HRP-DCM
300	1.0997	1.0006	0.9694	0.9017
600	2.3870	1.9403	1.2599	1.0579
1,200	5.6396	4.0945	3.8692	3.0165
2,400	7.4373	7.0467	6.0924	4.9201
3,000	8.4678	8.0194	7.1961	6.282
3,600	11.5937	10.1638	10.4981	8.8092

Table 7.2. *Energy dissipated in a network with 50 nodes*

Other results obtained for a WSN with 500 NCs show that the HRP-DCM solution optimizes the energy consumption best (Table 7.3).

The analysis of the results of Figure 7.4 and Tables 7.2 and 7.3 shows that the HRP-DCM approach consumes less energy than the competing approaches, and therefore, extends the lifetime of the WSN. Generally, the denser the network, the greater the number of clusters. Moreover, the greater the number of clusters, the better the network is covered. As shown in the results of Figure 7.5, implementation of the HRP-DCM solution, in the majority of cases, generates almost as many clusters as the HHRP solution, and therefore, ensures better coverage. This allows distant sensor nodes or those located on the edge of clusters to communicate directly with the respective CHs.

Execution time [s]	Total energy consumed [J]			
	LEACH	TEEN	HHRP	HRP-DCM
300	1.1206	1.0743	1.0043	0.8945
600	2.8285	2.5836	2.1964	1.9264
1,200	3.1543	3.0819	2.8937	2.5930
2,400	6.1490	5.8276	4.2834	3.9201
3,000	9.0874	7.7253	5.5398	4.7234
3,600	10.8054	9.0372	7.4310	5.8119
7,200	12.5142	11.1037	9.0920	7.1422
9,000	13.9854	12.4935	10.7492	9.0153
15,000	15.1659	13.7381	11.3198	10.3000
20,000	16.6198	15.2463	12.2391	11.9183
30,000	18.7514	16.2915	13.6346	12.2091

Table 7.3. *Energy dissipated in a network with 500 nodes*

For the same density, the HRP-DCM solution ensures better coverage of the network, leading all sensor nodes and CHs to transmit their information over short distances and therefore reduce the amount of energy related to the radio. For instance, in a WSN with a density of 200 nodes, the process begins with 23 and 21 clusters respectively for HRP-DCM and HHRP, and maintains this difference until all nodes have been exhausted. In this example, the HRP-DCM solution gains over 10 rounds in the lifetime of the network, and it reaches over 13 rounds for a network with 500 nodes.

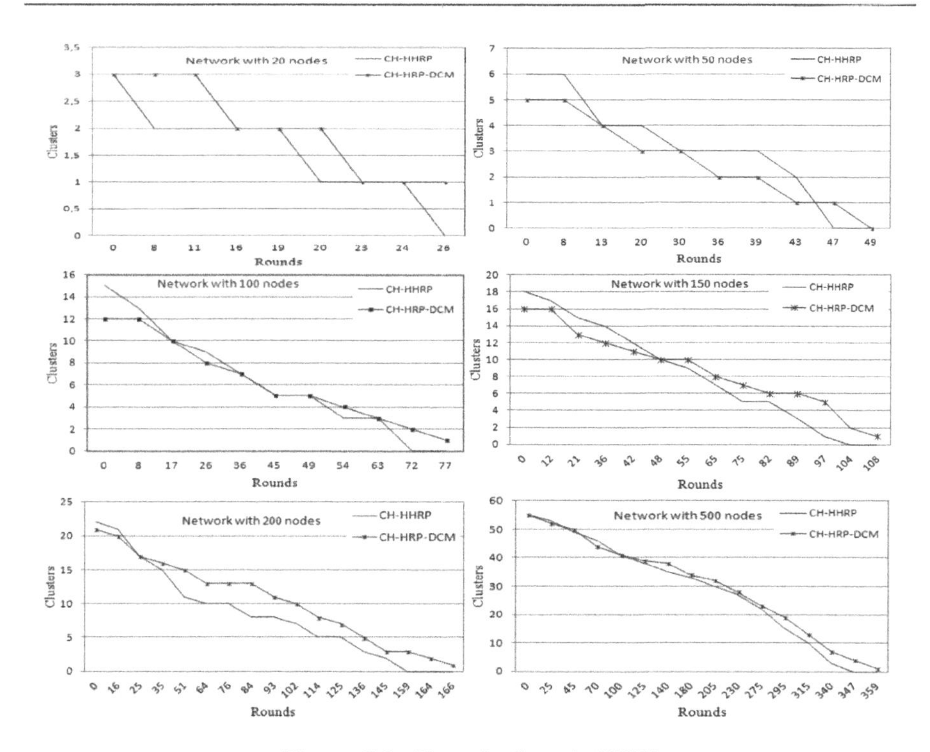

Figure 7.5. *Clusterization of a WSN*

The traffic management in the network is largely controlled by two mechanisms that allow access to the MAC layer, namely the CSMA/CA and TDMA methods, in order to avoid potential collisions and interference during information exchanges, as well as congestion in different sensor nodes, particularly at the level of the CH coordinator nodes where a CH_{PARENT} can receive information about several CH_{CHILD}. Moreover, optimizing the number of messages exchanged between different CH_{CHILD} and CH_{PARENT} avoids the possibility of this reducing the energy consumed. Figure 7.6 shows the useful messages exchanged in the different networks under consideration, i.e. 50, 100, 150, 200 and 500 node networks. The messages related to the environmental recognition and cluster formation stages are not taken into account.

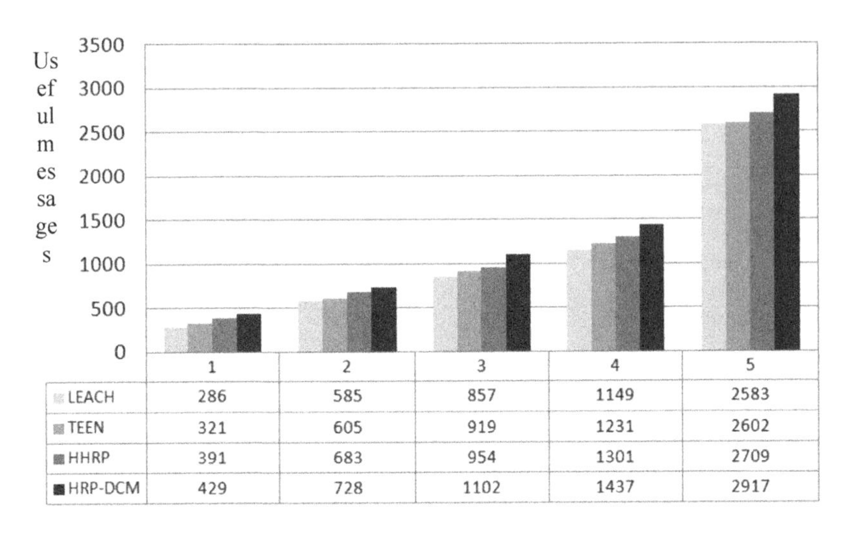

	1	2	3	4	5
LEACH	286	585	857	1149	2583
TEEN	321	605	919	1231	2602
HHRP	391	683	954	1301	2709
HRP-DCM	429	728	1102	1437	2917

Figure 7.6. *Network traffic. For a color version of the figure, see www.iste.co.uk/touati/energymanagement.zip*

We can see that the greater the density of the network, the greater the amount of traffic, which is completely normal. In this way, each network has a greater amount of clusters, and therefore, more intra- and inter-CH exchanges. We can also see that the two HRP-DCM and HHRP solutions use more messages. Indeed, given that the longevity of the network is greater than that in TEEN and LEACH, the two protocol solutions proposed have more operating time, and therefore, more message exchanges. The deployment of a WSN with 500 nodes exchanges more than 2,917 useful messages with HRP-DCM, but only 2,709 for HHRP, and 2,602 and 2,583 messages respectively for TEEN and LEACH.

Once the WSN has been deployed, the initialization stage requires significant amounts of energy to begin network operations by exchanging information between all sensor nodes and the base station. First, there is a stage of announcing a new round r, where the base station broadcasts a probability $P(t)$ throughout the network, leading each sensor node to make a decision on its future status by, in turn, generating a probability Pi to become CH. Once it has been chosen, each CH informs its surroundings of its new state by diffusing an ADV warning message containing its identifier. In order to do this, the MAC CSMA protocol is used to avoid collisions between different CHs. The decision to belong to a CH is based on the received signal strength, in which the CH node with the strongest signal will be chosen.

Once the clusters have been formed, each CH node ensures the coordination of data transmission in its own group. It uses a communication model, which transmits to all of its MNs via a TDMA exchange protocol. In this way, each MN has a time slot that allows it to be active over a time period allotted to it by its CH and to communicate with it, and it is in an inactive mode for the remainder of the time, which avoids it being in surveillance or passive listening modes, and thereby, minimizes energy consumption. In HRP-DCM solution, we have implemented the concept of calculating temporal distances in order to optimize the number of intra- and inter-cluster communications exchanged and avoid using the RSSI signal strength during cluster creation. In fact, using RSSI measures at all levels of the mechanism can cause deterioration in the network due to the asymmetry of communication links and waiting times for stabilizing the same important measures required for use. The results of Figure 7.7 show the performance of the new HRP-DCM solution in comparison to HHRP and LEACH.

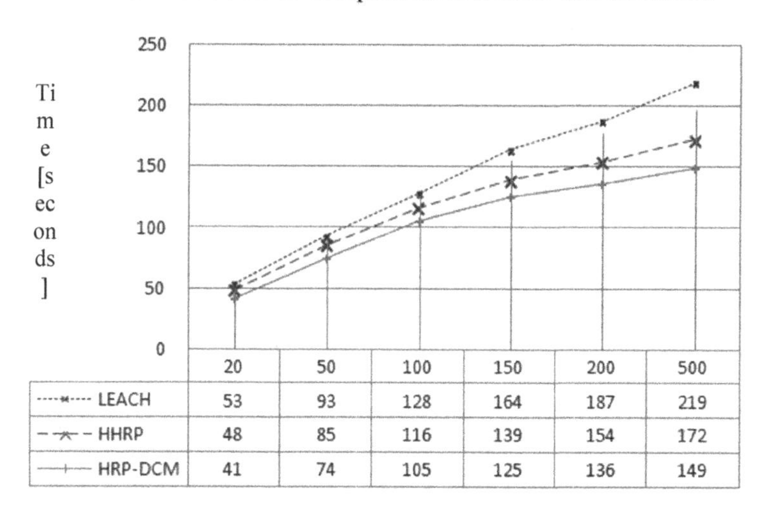

	20	50	100	150	200	500
LEACH	53	93	128	164	187	219
HHRP	48	85	116	139	154	172
HRP-DCM	41	74	105	125	136	149

Figure 7.7. *Environmental recognition and clusterization time delay. For a color version of the figure, see www.iste.co.uk/touati/energymanagement.zip*

We can clearly see that the delays concerning environmental recognition and clusterization increase in accordance with the density of the network. The larger the network, the greater the time spent hierarchizing nodes and forming clusters during initialization. During implementation of the HRP-DCM solution, a WSN with 100 nodes takes exactly 105 seconds to form clusters, while for a network with 500 nodes the time is 149 seconds, with a 44-second gap, which is normal. For the same 100 and 500 node networks, the times allotted for the HHRP solution are 116 and 172 seconds respectively. It is clear that, regardless of the network density, the HRP-DCM solution is better than HHRP and LEACH. The average gains in terms of time for environmental recognition and clusterization for a 50 node network are 13% and 21% respectively in contrast to HHRP and LEACH. It is 14% and 32% respectively compared with HHRP and LEACH for a network 10 times larger. Significant timeframes cause an overconsumption of energy, accelerating the loss of nodes and of the network as a whole.

7.5. Conclusion

In this chapter, we carried out a comparative study of different dedicated routing protocols for WSNs with a hierarchical architecture, such as HRP-DCM, HHRP, TEEN and LEACH. For this purpose, we focused on several metrics: energy consumption, end-to-end information conveyance times and the number of messages exchanged.

Implementation of the HRP-DCM solution bypasses the need to use RSSI measures at all levels of the mechanism, i.e. environmental recognition, clusterization and the construction of optimal paths according to the child–parent concept. The RSSI data can indeed be the cause of deterioration in the network due to asymmetry in communication links, causing a slowdown in data use as the RSSI signal is slow to stabilize.

The HRP-DCM solution is based on using the concept of calculating temporal distances during network deployment and environmental recognition. This concept allows the allotted times to be shortened for the initialization stage, thereby reducing the energy consumption.

The performance evaluation clearly shows that HRP-DCM is better at optimizing network operations, regardless of the network's density, in comparison to the HHRP solution and the TEEN and LEACH protocols.

Conclusion and Outlooks

In this book, we have addressed various issues related to WSN applications, particularly those connected to routing information. We have explored several solutions that have been developed in this regard by focusing on the factors that improve and/or worsen the performance and functioning of networks. This study has shown us that, despite significant advances in the miniaturization and integration of electronic components, the implementation of these solutions must take into consideration functional and/or structural constraints, such as the inherent features of sensors (energy consumption, calculation and memory), as well as environmental features (network topology, lack of infrastructure and loss of nodes). In fact, deploying a WSN in a hostile environment makes it vulnerable to multiple breakdowns ranging from physical defects caused by environmental factors to a lack of energy resources as a result of battery exhaustion. Human intervention is generally difficult or almost impossible to carry out due to the placement of sensors. As a result, management of energy consumption becomes an unavoidable issue in designing and implementing a WSN. Moreover, a guarantee of efficient functioning with increased network longevity will not be achieved by seeking only a material solution. It is therefore necessary to turn to other software solutions that will allow information use to be controlled from its source to its conveyance towards its final destination, taking the inherent features of sensors into account. Partially responding to these needs involves the development of computing tools and low-energy protocol strategies that implement mechanisms based on information routing techniques.

In order to do this, we have shown the work of two protocol solutions that deal with WSNs with a hierarchical architecture: the inheritance-based HHRP and HRP-DCM. The first solution is based on a clusterization mechanism and a nonlinear energy model that considers three sources of energy consumption, namely radio, processing and cluster formation. The mechanism uses the child–parent communication concept and selects the best paths on the basis of minimizing performance functions in critical parameters: the distance between source and target, the energy consumption and the signal strength.

Its implementation has three main stages: a stage of network deployment, initialization and recognition in which sensor nodes with well-defined features that are specific to the type of application are deployed at random in the operational environment, a dynamic clusterization stage that considers the different states of the sensor nodes, and a communication stage in which a single or multi-hop child–parent communication is proposed to optimize paths and packet transportation times between source and destination. In contrast to the standard LEACH protocol, where the nodes repeatedly select a cluster to which to belong, in HHRP the node selects the CH with the most energy for sending a single message to become an MN. This procedure for selecting new coordinator nodes does not pass through the base station, but occurs directly at the level of each cluster. The outgoing CH selects a new CH at the end of each round in accordance primarily with its energy level, and, if not, according to the received signal strength, if required.

The disadvantage we have noticed is that, during the environmental recognition stage, the inheritance-based solution requires the strength of RSSI radio signals to be used to estimate the distances between different sensor nodes. However, in WSNs, the communication links are asymmetrical, and given that the strength of radio signals is measured upon reception, this can cause deterioration in the performance of the proposed routing mechanism and errors in estimating the RSSI values. These values can only be used to locate nodes after a certain amount of time as the signal stabilizes slowly. In this way, the time allotted for the recognition stage becomes significant, accelerating energy consumption and thereby minimizing the lifetime of the network. In order to resolve this, we propose an alternative protocol based on hybrid routing with a dynamic HRP-DCM clusterization mechanism. Implementing this solution avoids the use of RSSI measures at all levels of the mechanism, i.e. not only during the environmental recognition and clusterization stages, but also during the communication stage to build optimal paths according to the child–parent concept. The HRP-DCM solution is based on a temporal distance calculation concept.

The performance evaluation clearly shows that HRP-DCM optimizes the network operations best, regardless of the density, in comparison to inheritance-based solutions and the TEEN and LEACH routing protocols. In execution times ranging from 300 to 3,600 s, HRP-DCM allows energy consumption to be optimized by an average gain factor equivalent to 10%. Implementing the HRP-DCM solution generates, in the majority of cases with a few exceptions, almost as many clusters as all other solutions, thereby ensuring better coverage. This allows distant nodes or those located on the edge of clusters to communicate directly or in a minimum number of hops with the respective CHs, thereby reducing the amount of energy related to the radio.

Moving away from the results obtained, we can extract a number of research perspectives from both a theoretical and applied point of view. In this way, investigations can focus on potential extensions of the protocol solutions proposed for routing information, taking into consideration the deployment environment and the deployment itself, as well as heterogeneity among the components used. We can first note that even if techniques based on the mobility of sensor nodes, such as the conveyance of messages by MULE agents, present the network with limitations in terms of execution time and processing, particularly when exchanges by appointment are not guaranteed, it would be interesting to look at the development of hybrid approaches that ensured coordination between different parameters, such as the positioning and monitoring of sensor nodes over time, the energy consumption, and the management of inter- and intra-cluster priorities. The use of rapid prototyping algorithms, such as those developed in [TOU 10, TOU 11a], would allow the clusterization process to accelerate, given that convergence is guaranteed after a maximum of three iterations, i.e. after several milliseconds. Moreover, we can have a process that can be fully interpreted by fuzzy linguistic models and rules.

From an application point of view, it would be very interesting to turn to other new technologies implementing WBAN networks, for example for embedded applications related to health, such as those developed within the framework of our research work on assisting persons with limited mobility [TOU 11b, TOU 11c, TOU 13] and the interpretation of thoughts using wireless sensor devices.

Bibliography

[AKY 02] AKYILDIZ I.F., SU W., SANKARASUBRAMANIAM Y. *et al.*, "A survey on sensor networks", *IEEE Communications Magazine*, vol. 40, no. 8, pp. 102–114, 2002.

[AKY 05] AKYILDIZ I.F., POMPILI D., MELODIA T., "Underwater acoustic sensor networks: research challenges", *Journal of Ad Hoc Networks*, vol. 3, no. 3, pp. 257–279, 2005.

[ALK 04] AL-KARAKI J.N., KAMAL A.E., "Routing techniques in wireless sensor networks: a survey", *IEEE Wireless Communications*, vol. 11, pp. 6–28, 2004.

[ANA 09] ANASTASI G., CONTI M., DI FRANCESCO M. *et al.*, "Energy conservation in wireless sensor networks: a survey", *Ad Hoc Networks*, vol. 7, no. 3, pp. 537–568, 2009.

[ANS 08] ANSARI J., PANKIN D., MAHONEN P., "Radio-triggered wake-ups with addressing capabilities for extremely low power sensor network applications", *Proceedings of the 19th International Symposium on Personal, Indoor and Mobile Radio Communications*, Cannes, France, pp. 1–5, 2008.

[AOU 12a] AOUDIA H., TOUATI Y., ALI-CHERIF A. *et al.*, "Hierarchical routing approach-based energy optimization in wireless sensor networks", *The 10th ACM International Symposium on Mobility Management and Wireless Access, MOBIWAC'12*, Paphos, Cyprus, pp. 131–134, October 2012.

[AOU 12b] AOUDIA H., TOUATI Y., GREUSSAY P. *et al.*, "Energy optimization in wireless sensor networks", *The 11th International Conference on Wireless Networks, World Congress in Computer Science, Computer Engineering and applied Computing, ICWN'12*, Las Vegas, NV, pp. 470–475, July 2012.

[AOU 13a] AOUDIA H., TOUATI Y., ALI-CHERIF A., "Energy-efficient routing protocol based on a dynamic clustering mechanism for WSN applications", *The 11th ACM International Symposium on Mobility Management and Wireless Access*, Barcelona, Spain, pp. 89–92, November 2013.

[AOU 13b] AOUDIA H., TOUATI Y., ALI-CHERIF A., "Energy optimization mechanism in wireless sensor networks", *MobilWare 2013*, Bologna, Italy, pp. 94–99, December 2013.

[AOU 14a] AOUDIA H., TOUATI Y., TEGUIG E.H. *et al.*, "Hybrid hierarchical routing protocol for WSN lifetime maximization", *International Conference on Wireless Communications and Applications*, Barcelona, Spain, pp. 928–934, October 2014.

[AOU 14b] AOUDIA H., TOUATI Y., ALI-CHERIF A., "Wireless sensor networks lifetime extension based on a new hybrid hierarchical routing mechanism", *The 8th International Wireless Internet Conference – Symposium on Wireless and Vehicular Communication*, Lisbon, Portugal, pp. 262–270, November 2014.

[AOU 15a] AOUDIA H., TOUATI Y., ALI-CHERIF A., "Wireless sensor networks lifetime extension based on a new hybrid hierarchical routing mechanism", *Wireless Internet*, Springer, vol. 146, pp. 262–270, 2015

[AOU 15b] AOUDIA H., Adaptive routing approach for optimizing energy consumption in WSN applications, Ph.D. Thesis, University of Paris 8, December 2015.

[ARI 02] ARISHA K., YOUSSEF M.A., YOUNIS M.F., "Energy-aware TDMA-based MAC for sensor networks", in KARRY R., GOODMAN D. (ed.), *System-Level Power Optimization for Wireless Multimedia Communication*, Kluwer Academic, 2002.

[ASL 02] ASLAM J., LI Q., RUS D., "Three power-aware routing algorithms for sensor networks", *Journal of Wireless Communications Mobile Computing*, vol. 3, pp. 187–208, 2002.

[BAL 11] BALAMURUGAN P., DURAISWAMY K., "Consistent and proficient algorithm for data gathering in wireless sensor networks", *Journal of Computer Science*, vol. 7, no. 9, pp. 1400–1406, 2011.

[BAS 07] BASAGNI S., CAROSI A., MELACHRINOUDIS E. *et al.*, "Controlled sink mobility for prolonging wireless sensor networks lifetime", *Journal of Wireless Networks*, vol. 14, no. 6, pp. 831–858, 2007.

[BRA 02] BRAGINSKY D., ESTRIN D., "Rumor routing algorithm for sensor networks", *Proceedings of the 1st ACM International Workshop on Wireless Sensor Networks and Applications*, Atlanta, GA, pp. 22–31, 2002.

[BUL 00] BULUSU N., HEIDEMANN J., ESTRIN D., GPS-less low cost outdoor localization for very small devices, Technical Report 00-729, Computer Science Department, University of Southern California, April 2000.

[CAP 01] CAPKUN S., HAMDI M., HUBAUX J., "GPS-free positioning in mobile ad-hoc networks", *Proceedings of the 34th Annual Hawaii International Conference on System Sciences*, pp. 3481–3490, 2001.

[CAS 05] CASARI P., MARCUCCI A., NATI M. *et al.*, "A detailed simulation study of geographic random forwarding (GeRaF) in wireless sensor networks", *Proceedings of IEEE Military Communications Conference*, vol. 1, pp. 59–68, 2005.

[CER 04] CERPA A., ESTRIN D., "ASCENT: adaptive self-configuring sensor network topologies", *IEEE Transactions on Mobile Computing*, vol. 3, no. 3, pp. 272–285, 2004.

[CHA 04] CHANG J.H., TASSIULAS L., "Maximum lifetime routing in wireless sensor networks", *IEEE/ACM Transactions on Networking*, vol. 12, no. 4, pp. 609–619, 2004.

[CHE 02] CHEN B., JAMIESON K., BALAKRISHNAN H. *et al.*, "SPAN: an energy-efficient coordination algorithm for topology maintenance in ad hoc wireless networks", *Wireless Networks Journal*, vol. 8, no. 5, pp. 481–494, 2002.

[CHE 06] CHEN H., MEGERIAN S., "Cluster sizing and head selection for efficient data aggregation and routing in sensor networks", *IEEE Conference on Wireless Communications and Networking*, Las Vegas, NV, pp. 2318–2323, 2006.

[CHO 03] CHONG C.Y., KUMAR S., "Sensor networks: evolution, opportunities, and challenges", *Proceedings of the IEEE*, vol. 91, no. 8, pp. 1247–1256, 2003.

[CHU 02] CHU M., HAUSSECKER H., ZHAO F., "Scalable information-driven sensor querying and routing for ad hoc heterogeneous sensor networks", *International Journal of High Performance Computing Applications*, vol. 16, no. 3, pp. 293–313, 2002.

[CHU 06] CHU D., DESHPANDE A., HELLERSTEIN J.M. *et al.*, "Approximate data collection in sensor networks using probabilistic models", *Proceedings of the 22nd International Conference on Data Engineering*, Atlanta, GA, pp. 48–60, April 2006.

[DAM 03] DAM T.V., LANGENDOEN K., "An adaptive energy-efficient MAC Protocol for wireless sensor networks", *Proceedings of the 1st International Conference on Embedded Networked Sensor Systems*, Los Angeles, CA, pp. 171–180, November 2003.

[DEM 06] DEMIRKOL I., ERSOY C., ALAGOZ F., "MAC protocols for wireless sensor networks: a survey", *IEEE Communications Magazine*, vol. 44, no. 4, pp. 115–121, 2006.

[DES 04] DESHPANDE A., GUESTRIN C., MADDEN S. *et al.*, "Model-driven data acquisition in sensor networks", *Proceedings of the 30th International Conference on Very Large Data Bases*, Toronto, Canada, vol. 30, pp. 588–599, 2004.

[DUL 03] DULMAN S., NIEBERG T., WU J. *et al.*, "Trade-off between traffic overhead and reliability in multipath routing for wireless sensor networks", *IEEE Wireless Communications and Networking*, New Orleans, LO, vol. 3, pp. 1918–1922, March 2003.

[EPH 82] EPHREMIDES A., MOWAFI O., "Analysis of a hybrid access scheme for buffered user-probabilistic time division", *IEEE Transactions on Software Engineering*, vol. 8, no. 1, pp. 52–61, 1982.

[ERR 08] ERRAMILLI V., CROVELLA M., "Mobile-relay forwarding in opportunistic networks with resource constraints", *Proceedings of the 3rd ACM Workshop on Challenged Networks*, pp. 41–48, 2008.

[FAS 07] FASOLO E., ROSSI M., WIDMER J. *et al.*, "In-network aggregation techniques for wireless sensor networks: a survey", *IEEE Wireless Communications*, vol. 14, no. 2, pp. 70–87, 2007.

[FRI 46] FRIIS H.T., "A note on a simple transmission formula", *Proceedings of the IRE*, vol. 34, no. 5, pp. 254–256, 1946.

[GAN 03] GANDHAM S.R., DAWANDE M., PRAKASH R. *et al.*, "Energy efficient schemes for wireless sensor networks with multiple mobile base stations", *Proceedings of the IEEE Global Communications Conference, GLOBECOM*, San Francisco, CA, vol. 1, pp. 377–381, 2003.

[GAN 04] GANESAN D., CERPA A., YE W. *et al.*, "Networking issues in wireless sensor networks", *Journal of Parallel and Distributed Computing*, vol. 64, pp. 799–814, 2004.

[GAY 03] GAY D., LEVIS P., VON-BEHREN R. *et al.*, "The NesC language: a holistic approach to networked embedded systems", *ACM SIGPLAN Conference on Programming Language Design and Implementation*, San Diego, CA, pp. 1–11, June 2003.

[GOD 04] GODFREY P.B., RATAJCZAK D., "Naps: scalable, robust topology management in wireless ad hoc networks", *Proceedings of the 3rd International Symposium on Information Processing in Sensor Networks*, Berkeley, CA, pp. 443–451, 2004.

[GOE 06] GOEL S., PASSARELLA A., IMIELINSKI T., "Using buddies to live longer in a boring world", *Proceedings of the 4th Annual IEEE International Conference on Pervasive Computing and Communications Workshops*, Pisa, Italy, pp. 342–347, 2006.

[GU 05] GU L., STANKOVIC J., "Radio-triggered wake-up for wireless sensor networks", *Real-Time Systems Journal*, vol. 29, pp. 157–182, 2005.

[HAA 00] HAARTSEN J.C., "The bluetooth radio system", *IEEE Personal Communications*, vol. 7, no. 1, pp. 28–36, 2000.

[HAA 06] HAAS Z.J., SMALL T., "A new networking model for biological applications of ad hoc sensor networks", *IEEE/ACM Transactions on Networking*, vol. 14, no. 1, pp. 27–40, 2006.

[HAL 09] HALGAMUGE M.N., ZUKERMAN M., RAMAMOHANARAO K., "An estimation of sensor energy consumption", *Progress in Electromagnetics Research B*, vol. 12, pp. 259–295, 2009.

[HAN 04] HAN Q., MEHROTRA S., VENKATASUBRAMANIAN N., "Energy efficient data collection in distributed sensor environments", *Proceedings of the 24th IEEE International Conference on Distributed Computing Systems*, pp. 590–597, 2004.

[HE 03] HE T., STANKOVIC J.A., CHENYANG L. *et al.*, "SPEED: a stateless protocol for real-time communication in sensor networks", *Proceedings of the 23rd International Conference on Distributed Computing Systems*, Providence, RI, pp. 46–55, May 2003.

[HED 88] HEDETNIEMI S., LIESTMAN A., "A survey of gossiping and broadcasting in communication networks", *IEEE Networks*, vol. 18, no. 4, pp. 319–349, 1988.

[HEI 99] HEINZELMAN W., KULIK J., BALAKRISHNAN H., "Adaptive protocols for information dissemination in wireless sensor networks", *Proceedings of the 5th ACM/IEEE Mobicom Conference*, Seattle, WA, pp. 174–85, 1999.

[HEI 00] HEINZELMAN W., CHANDRAKASAN A., BALAKRISHNAN H., "Energy-efficient communication protocol for wireless microsensor networks", *Proceedings of the 33rd Hawaii International Conference on System Sciences*, vol. 8, pp. 8020–8030, 2000.

[HEI 02] HEINZELMAN W.R., CHANDRAKASAN A., BALAKRISHNAN H., "An application-specific protocol architecture for wireless micro-sensor networks", *IEEE Transactions on Wireless Communications*, vol. 1, no. 4, pp. 660–670, 2002.

[HIL 00] HILL J., SZEWCZYK R., WOO A. *et al.*, "System architecture directions for networked sensors", *ACM SIGPLAN*, vol. 35, no. 11, pp. 93–104, 2000.

[INT 00] INTANAGONWIWAT C., GOVINDAN R., ESTRIN D., "Directed diffusion: a scalable and robust communication paradigm for sensor networks", *Proceedings of the 6th Annual ACM/IEEE International Conference on Mobile Computing and Networking*, Boston, MA, pp. 56–67, August 2000.

[INT 03] INTANAGONWIWAT C., GOVINDAN R., ESTRIN D. *et al.*, "Directed diffusion for wireless sensor networking", *IEEE/ACM Transactions on Networking*, vol. 11, no. 1, pp. 2–16, 2003.

[JAI 06] JAIN S., SHAH R.C., BRUNETTE W. *et al.*, "Exploiting mobility for energy efficient data collection in wireless sensor networks", *Journal of Mobile Networks and Applications*, vol. 11, no. 3, pp. 327–339, 2006.

[JEA 05] JEA D., SOMASUNDARA A.A., SRIVASTAVA M.B., "Multiple controlled mobile elements (data mules) for data collection in sensor networks", *Proceedings of the 1st IEEE/ACM International Conference on Distributed Computing in Sensor Systems*, Marina del Rey, CA, pp. 244–255, June 2005.

[JOH 96] JOHNSON D.B., MALTZ D.A., "Dynamic source routing in ad hoc wireless networks", in IMIELINSKI T., KORTH H. (eds), *Mobile Computing*, Springer, 1996.

[JOH 01] JOHNSON D.B., MALTZ D.A., BROCH J., "DSR: the dynamic source routing protocol for multi-hop wireless ad hoc networks", in PERKINS C.E. (ed.), *Ad Hoc Networking*, Addison-Wesley, 2001.

[JUA 02] JUANG P., OKI H., WANG Y. *et al.*, "Energy-efficient computing for wildlife tracking: design tradeoffs and early experiences with Zebranet", *Proceedings of the 10th International Conference on Architectural Support for Programming Languages and Operating Systems*, pp. 94–107, 2002.

[JUN 05a] JUN H., AMMAR M.H., ZEGURA E.W., "Power management in delay tolerant networks: a framework and knowledge-based mechanisms", *Proceedings of the IEEE Conference on Sensor and Ad Hoc Communications and Networks*, pp. 418–429, 2005.

[JUN 05b] JUN H., ZHAO W., AMMAR M. *et al.*, "Trading latency for energy in wireless ad hoc networks using message ferrying", *Proceedings of the 3rd IEEE International Conference on Pervasive Computing and Communications Workshops*, pp. 220–225, March 2005.

[JUN 07] JUNG S.M., HAN Y.J., CHUNG T.M., "The concentric clustering scheme for efficient energy consumption in PEGASIS", *9th International Conference on Advanced Communication Technology*, vol. 1, pp. 260–265, 2007.

[KAH 99] KAHN J.M., KATZ R.H., PISTER K.S.J., "Next century challenges: mobile networking for smart dust", *Proceedings of the ACM MobiCom'99*, Seattle, WA, pp. 271–278, 1999.

[KAN 04] KANSAL A., SOMASUNDARA A.A., JEA D.D. *et al.*, "Intelligent fluid infrastructure for embedded networks", *Proceedings of the 2nd ACM International Conference on Mobile Systems, Applications, and Services*, Boston, MA, pp. 111–124, June 2004.

[KAN 08] KANAGAL B., DESHPANDE A., "Online filtering, smoothing and probabilistic modeling of streaming data", *Proceedings of the 24th International Conference on Data Engineering*, Cancun, Mexico, pp. 1160–1169, 7–12 April 2008.

[KAR 00] KARP B., KUNG H.T., "GPSR: greedy perimeter stateless routing for wireless sensor networks", *Proceedings of the 6th Annual ACM/IEEE International Conference on Mobile Computing and Networking*, Boston, MA, pp. 243–254, August 2000.

[KES 06] KESHAVARZIAN A., LEE H., VENKATRAMAN L., "Wakeup scheduling in wireless sensor networks", *Proceedings of the ACM MobiHoc*, Florence, Italy, pp. 322–333, 2006.

[KHA 10] KHAN A., ABDULLAH A., HASAN N., "Maximizing lifetime of homogeneous WSN through energy efficient clustering method", *International Journal of Computer Science and Security*, vol. 3, no. 6, pp. 583–594, 2010.

[KON 07] KONG Z., YEH E.M., "Distributed energy management algorithm for large-scale wireless sensor networks", *Proceedings of the 8th ACM International Symposium on Mobile Ad Hoc Networking and Computing*, Montreal, Canada, pp. 209–218, 2007.

[KUH 03] KUHN F., WATTENHOFER R., ZOLLINGER A., "Worst-case optimal and average-case efficient geometric ad-hoc routing", *Proceedings of the 4th ACM International Conference on Mobile Computing and Networking*, pp. 267–278, 2003.

[KUL 02] KULIK J., HEINZELMAN W.R., BALAKRISHNAN H., "Negotiation-based protocols for disseminating information in wireless sensor networks", *Wireless Networks*, vol. 8, pp. 169–185, 2002.

[KUM 11] KUMAR D., ASERI T.C., PATEL R.B., "EECDA: energy efficient clustering and data aggregation protocol for heterogeneous wireless sensor networks", *International Journal of Computers, Communications and Control*, vol. 6, no. 1, pp. 113–124, 2011.

[KUR 07] KURTIS K., PRASANT M., "Medium access control in wireless sensor networks", *Computer Networks*, vol. 51, no. 4, pp. 961–994, 2007.

[LAN 03] LANGENDOEN K., REIJERS N., "Distributed localization in wireless sensor networks: a quantitative comparison", *Computer Networks*, vol. 43, no. 4, pp. 499–518, 2003.

[LE 07] LE BORGNE Y.A., SANTINI S., BONTEMPI G., "Adaptive model selection for time series prediction in wireless sensor networks", *Journal of Signal Processing*, vol. 87, no. 12, pp. 3010–3020, 2007.

[LEV 03] LEVIS P., LEE N., WELSH M. *et al.*, "TOSSIM: accurate and scalable simulation of entire TinyOS applications", *Proceedings of the 1st ACM Conference on Embedded Networked Sensor Systems*, New York, NY, pp. 126–37, 2003.

[LEV 04] LEVIS P., MADDEN S., GAY D. *et al.*, "The emergence of networking abstractions and techniques in TinyOS", *Proceedings of the 1st USENIX/ACM Symposium on Networked Systems Design and Implementation*, Berkeley, CA, vol. 1, pp. 1–14, 2004.

[LEV 09] LEVIS P., GAY D., *TinyOS Programming*, 1st ed., Cambridge University Press, 2009.

[LI 01a] LI Q., ASLAM J., RUS D., "Online power-aware routing in sensor networks", *Proceedings of the 7th Annual International Conference on Mobile Computing and Networking*, Rome, Italy, pp. 97–107, May 2001.

[LI 01b] LI L., HALPERN J.Y., "Minimum-energy mobile wireless networks revisited", *Proceedings of IEEE International Conference on Communications*, Helsinki, Finland, vol. 1, pp. 278–283, July 2001.

[LI 04] LI J., LAZAROU G., "A bit-map-assisted energy-efficient MAC scheme for wireless sensor networks", *Proceedings of the 3rd International Symposium on Information Processing in Sensor Networks*, Berkeley, CA, pp. 55–60, April 2004.

[LI 07] LI J., MOHAPATRA P., "Analytical modeling and mitigation techniques for the energy hole problem in sensor networks", *Pervasive Mobile Computing*, vol. 3, no. 3, pp. 233–254, 2007.

[LI 11] LI C., ZHANG H., HAO B. *et al.*, "A survey on routing protocols for large-scale wireless sensor networks", *Sensors*, vol. 11, pp. 3498–3526, 2011.

[LIA 05] LIANG Y., YU H., "Energy adaptive Cluster-Head selection for wireless sensor networks", *Proceedings of the 6th International Conference on Parallel and Distributed Computing, Applications and Technologies, PDCAT*, pp. 634–638, 2005.

[LIN 02] LINDSEY S., RAGHAVENDRA C., "PEGASIS: power-efficient gathering in sensor information systems", *Proceedings of IEEE Aerospace Conference*, vol. 3, nos. 9–16, pp. 1125–1130, 2002.

[LOS 05] LOSCRI V., MORABITO G., MARANO S., "A two-level hierarchy for low-energy adaptive clustering hierarchy (TL-LEACH)", *Proceedings of the 62nd IEEE Vehicular Technology Conference*, vol. 3, pp. 1809–1813, September 2005.

[LU 04] LU G., KRISHNAMACHARI B., RAGHAVENDRA C.S., "An adaptive energy-efficient and low-latency mac for data gathering in wireless sensor networks", *Proceedings of the 18th International Symposium on Parallel and Distributed Processing*, April 2004.

[MAI 02] MAINWARING A., POLASTRE J., SZEWCZYK R. *et al.*, "Wireless sensor networks for habitat monitoring", *Proceedings of the ACM Workshop on Wireless Sensor Networks and Applications*, Atlanta, GA, pp. 88–97, September 2002.

[MAN 01] MANJESHWAR A., AGARWAL D.P., "TEEN: a routing protocol for enhanced efficiency in wireless sensor networks", *Proceedings of the 15th International Symposium on Parallel and Distributed Computing*, San Francisco, CA, pp. 2009–2015, 2001.

[MAN 02] MANJESHWAR A., AGARWAL D.P., "APTEEN: a hybrid protocol for efficient routing and comprehensive information retrieval in wireless sensor networks", *Proceedings of International Parallel and Distributed Processing Symposium*, pp. 195–202, 2002.

[MEH 11] MEHRANI M., SHANBEHZADEH J., SARRAFZADEH A. *et al.*, "FEED: fault tolerant, energy-efficient, distributed clustering for WSN", *Global Journal of Computer Science and Technology*, vol. 11, no. 4, pp. 58–65, 2011.

[MHA 04] MHATRE V., ROSENBERG C., "Design guidelines for wireless sensor networks: communication, clustering and aggregation", *Ad Hoc Networks Journal*, vol. 2, no. 1, pp. 45–63, 2004.

[MIL 05] MILLER M.J., VAIDYA N.H., "A MAC protocol to reduce sensor network energy consumption using a wakeup radio", *IEEE Transactions on Mobile Computing*, vol. 4, no. 3, pp. 228–242, 2005.

[MIR 05] MIRZA D., OWRANG M., SCHURGERS C., "Energy-efficient wakeup scheduling for maximizing lifetime of IEEE 802.15.4 networks", *Proceedings of the International Conference on Wireless Internet*, Budapest, Hungary, pp. 130–137, 2005.

[MUN 09] MUNOZ D., BOUCHEREAU F., VARGAS C. *et al.*, *Position Location Techniques and Applications*, Elsevier, 2009.

[NAT 03] NATH B., NICULESCU D., "Routing on a curve", *ACM SIGCOMM Computer Communication Review*, vol. 33, no. 1, pp. 155–160, 2003.

[PAH 95] PAHLAVAN L., LEVESQUE A., *Wireless Information Network*, Wiley-InterScience, New York, 1995.

[PAO 07] PAOLO B., PRASHANT P., VINCE W.C., "Wireless sensor networks: a survey on the state of the art and the 802.15.4 ZigBee standards", *Computer Communications*, vol. 30, no. 7, pp. 1655–1695, 2007.

[PAP 06] PAPADIMITRIOU I., GEORGIADIS L., "Energy-aware routing to maximize lifetime in wireless sensor networks with mobile sink", *Journal of Communications Software and Systems*, vol. 2, no. 2, pp. 141–151, 2006.

[PER 99] PERKINS C.E., ROYER E.M., "Ad-hoc on-demand distance vector routing", *Proceedings of Mobile Computing Systems and Applications*, New Orleans, LA, pp. 90–100, 1999.

[PIR 11] PIRAN M.J., MURTHY G.R., BABU G.P., "Vehicular ad hoc and sensor networks: principles and challenges", *International Journal of Ad Hoc, Sensor and Ubiquitous Computing*, vol. 2, no. 2, pp. 38–49, 2011.

[POL 04] POLASTRE J., HILL J., CULLER D., "Versatile low power media access for sensor networks", *Proceedings of the 2nd ACM International Conference on Embedded Networked Sensor Systems*, pp. 95–107, 2004.

[POT 00] POTTIE G.J., KAISER W.J., "Wireless integrated network sensors", *ACM Communications*, vol. 43, no. 5, pp. 52–58, 2000.

[PRA 03] PRADHAN S.S., RAMCHANDRAN K., "Distributed source coding using syndromes DISCUS: design and construction", *IEEE Transactions on Information Theory*, vol. 49, no. 3, pp. 626–643, 2003.

[RAH 02] RAHUL C., RABAEY J., "Energy aware routing for low energy ad hoc sensor networks", *IEEE Wireless Communications and Networking Conference*, Orlando, FL, vol. 1, pp. 350–355, March 2002.

[RAJ 03] RAJENDRAN V., OBRACZA K., GARCIA-LUNA-ACEVES J.J., "Energy-efficient, collision-free medium access control for wireless sensor networks", *Proceedings of 1st International Conference on Embedded Networked Sensor Systems*, Los Angeles, CA, pp. 181–192, 2003.

[RAJ 05] RAJENDRAN V., OBRACZKA K., GARCIA-LUNA-ACEVES J.J., "Energy-efficient, application-aware medium access for sensor networks", *Proceedings of the 2nd IEEE International Conference on Mobile Ad-hoc and Sensor Systems*, Washington, DC, pp. 622–630, 2005.

[RAP 96] RAPPAPORT T.S., *Wireless Communications: Principles and Practice*, Prentice Hall, New Jersey, 1996.

[ROD 99] RODOPLU V., MENG T.H., "Minimum energy mobile wireless networks", *IEEE Journal on Selected Areas in Communications*, vol. 17, no. 8, pp. 1333–1344, 1999.

[ROY 10] ROYCHOWDHURY S., PATRA C., "Geographic adaptive fidelity and geographic energy aware routing in adhoc routing", *International Journal of Computer and Communication Technology*, vol. 2, nos. 2–4, pp. 309–313, 2010.

[SAD 03] SADAGOPAN N., KRISHNAMACHARI B., HELMY A., "The ACQUIRE mechanism for efficient querying in sensor networks", *Proceedings of the 1st International Workshop on Sensor Network Protocols and Applications*, Anchorage, AK, pp. 149–155, 2003.

[SAN 05] SANTI P., "Topology control in wireless adhoc and sensor networks", *ACM Computing Survey*, vol. 37, no. 2, pp. 164–194, 2005.

[SAV 01] SAVVIDES A., HAN C.-C., SRIVASTAVA M., "Dynamic fine-grained localization in adhoc networks of sensors", *Proceedings of the 7th ACM Annual International Conference on Mobile Computing and Networking*, pp. 66–179, July 2001.

[SCH 01] SCHURGERS C., SRIVASTAVA M.B., "Energy efficient routing in wireless sensor networks", *Proceedings of Military Communications Conference on Communications for Network-Centric Operations: Creating the Information Force*, McLean, VA, pp. 357–361, 2001.

[SEN 11] SEN F., BING Q., LIANGRUI T., "An improved energy-efficient PEGASIS-based protocol in wireless sensor networks", *8th International Conference on Fuzzy Systems and Knowledge Discovery*, vol. 4, pp. 2230–2233, 2011.

[SHA 02] SHAH R.C., RABAEY J., "Energy aware routing for low energy ad-hoc sensor networks", *IEEE Wireless Communications and Networking Conference*, Orlando, FL, pp. 17–21, 2002.

[SHA 03] SHAH R.C., ROY S., JAIN S. *et al.*, "Data MULEs: modeling a three-tier architecture for sparse sensor networks", *Proceedings of the 1st International Workshop on Sensor Network Protocols and Applications*, pp. 30–41, 2003.

[SHI 01] SHIH E., CHO S., ICKES N. *et al.*, "Physical layer driven protocol and algorithm design for energy-efficient wireless sensor networks", *Proceedings of ACM MobiCom*, Roma, Italy, pp. 272–286, 2001.

[SMA 03] SMALL T., HAAS Z., "The shared wireless infostation model: a new ad hoc networking paradigm (or where there is a whale, there is a way)", *Proceedings of the 4th ACM international Symposium on Mobile Ad Hoc Networking and Computing*, pp. 233–244, 2003.

[SOM 06] SOMASUNDARA A.A., KANSAL A., JEA D.D. *et al.*, "Controllably mobile infrastructure for low energy embedded networks", *IEEE Transactions on Mobile Computing*, vol. 5, no. 8, pp. 958–973, 2006.

[STO 99] STOJMENOVIC I., LIN X., "GEDIR: loop-free location based routing in wireless networks", *International Conference on Parallel and Distributed Computing and Systems*, Boston, MA, pp. 1025–1028, 1999.

[SUB 00] SUBRAMANIAN L., KATZ R.H., "An architecture for building self configurable systems", *Proceedings of IEEE/ACM Workshop on Mobile Ad Hoc Networking and Computing*, Boston, MA, pp. 63–73, 2000.

[TAN 04] TANG C., RAGHAVENDRA C.S., "Compression techniques for wireless sensor networks", *Wireless Sensor Networks*, Springer, Part 3, pp. 207–231, 2004.

[TIL 02] TILAK S., ABU-GHAZALEH N., HEINZELMAN W., "A taxonomy of wireless micro-senor network models", *ACM SIGMOBILE, Mobile Computing and Communications Review*, vol. 6, no. 2, pp. 28–36, 2002.

[TOU 10] TOUATI Y., AOUDIA H., ALI-CHERIF A., "Intelligent wheelchair localization in wireless sensor network environment: a fuzzy logic approach", *5th IEEE International Conference on Intelligent Systems*, London, UK, pp. 408–413, July 2010.

[TOU 11a] TOUATI Y., AOUDIA H., ALI-CHERIF A. *et al.*, "Position location technique in wireless sensor network using rapid prototyping algorithm", *Advanced Applications of Rapid Prototyping Technology in Modern Engineering*, INTECH Publisher, pp. 291–306, 2011.

[TOU 11b] TOUATI Y., ALI-CHERIF A., AOUDIA H. *et al.*, "Virtual Impedance approach for smart wheelchair monitoring via wireless communication network", *International Conference on Embedded Systems and Applications ESA'11, World Congress in Computer Science, Computer Engineering and applied Computing*, Las Vegas, NV, pp. 34–39, July 2011.

[TOU 11c] TOUATI Y., ALI-CHERIF A., AOUDIA H. *et al.*, "An embedded control architecture for smart wheelchair navigation via wireless network", *IEEE International Conference on Information Reuse and Integration*, Las Vegas, NV, pp. 492–493, August 2011.

[TOU 13] TOUATI Y., AOUDIA H., ALI-CHERIF A., "Virtual impedance approach for smart wheelchair monitoring via wireless communication network", *International Journal of Advanced Computer Science*, vol. 3, no. 4, pp. 175–183, 2013.

[TUB 03] TUBAISHAT M., MADRIA S., "Sensor networks: an overview", *IEEE Potentials*, vol. 22, no. 2, pp. 20–23, 2003.

[TUL 06a] TULONE D., MADDEN S., "PAQ: time series forecasting for approximate query answering in sensor networks", *Proceedings of the 3rd European Conference on Wireless Sensor Networks*, pp. 21–37, February 2006.

[TUL 06b] TULONE D., MADDEN S., "An energy-efficient querying framework in sensor networks for detecting node similarities", *Proceedings of the 9th International Symposium on Modeling, Analysis and Simulation of Wireless and Mobile Systems*, pp. 291–300, October 2006.

[ULE 06] ULEMA M., NOGUEIRA J.M., KOZBE B., "Management of wireless ad-hoc networks and wireless sensor networks", *Journal of Systems and Network Management, Springer*, vol. 14, no. 3, pp. 327–333, 2006.

[VAS 05] VASILESCU I., KOTAY K., RUS D. *et al.*, "Data collection, storage, and retrieval with an underwater sensor network", *Proceedings of the 3rd ACM International Conference on Embedded Networked Sensor Systems*, San Diego, CA, pp. 154–165, November 2005.

[VUR 04] VURAN M.C., AKAN O.B., AKYILDIZ I.F., "Spatio-temporal correlation: theory and applications for wireless sensor networks", *Computer Networks Journal*, vol. 45, no. 3, pp. 245–261, 2004.

[WAN 05] WANG Z.M., BASAGNI S., MELACHRINOUDIS E. *et al.*, "Exploiting sink mobility for maximizing sensor networks lifetime", *Proceedings of the 38th Annual Hawaii International Conference on System Sciences*, Hawaii, vol. 9, p. 287, 1 January 2005.

[WAR 07] WARRIER A., PARK S., MINA J. *et al.*, "How much energy saving does topology control offer for wireless sensor networks? A practical study", *ACM Computer Communications*, vol. 30, no. 14, pp. 2867–2879, 2007.

[WU 03] WU M., CHEN C.W., "Multiple bitstream image transmission over wireless sensor networks", *Proceedings of the IEEE Sensors*, vol. 2, pp. 721–731, 2003.

[XIO 04] XIONG Z., LIVERIS A.D., CHENG S., "Distributed source coding for sensor networks", *IEEE Signal Processing Magazine*, vol. 21, no. 5, pp. 80–94, 2004.

[XU 01] XU Y., HEIDEMANN J., ESTRIN D., "Geography-informed energy conservation for adhoc routing", *Proceedings of the 7th Annual International Conference on Mobile Computing and Networking*, Roma, Italy, pp. 70–84, 2001.

[YAD 09] YADAV R., VARMA S., MALAVIYA N., "Survey of MAC protocols for wireless sensor networks", *UbiCC Journal*, vol. 4, no. 3, pp. 827–834, 2009.

[YAN 07] YANG H., SIKDAR B., "Optimal cluster head selection in the LEACH architecture", *International Performance, Computing, and Communications Conference*, New Orleans, LO, pp. 93–100, 2007.

[YAO 02] YAO Y., GEHRKE J., "The cougar approach to in-network query processing in sensor networks", *ACM SIGMOD Record*, vol. 31, no. 3, pp. 9–18, 2002.

[YAS 09] YASSEIN M.B., AL-ZOUBI A., KHAMAYSEH Y. *et al.*, "Improvement on LEACH protocol of wireless sensor network", *Journal of Digital Content Technology and Its Applications*, vol. 3, no. 1, pp. 132–136, 2009.

[YE 00] YE W., HEIDERMANN J., ESTRIN D., "An energy-efficient MAC protocol for wireless sensor networks", *Proceedings of the IEEE INFOCOM*, vol. 3, pp. 1567–1576, June 2000.

[YE 01] YE F., CHEN A., LIU S. *et al.*, "A scalable solution to minimum cost forwarding in large sensor networks", *IEEE Proceedings of the 10th International Conference on Computer Communications and Networks*, pp. 304–309, 2001.

[YE 04] YE W., HEIDEMANN J., ESTRIN D., "Medium access control with coordinated adaptive sleeping for wireless sensor networks", *IEEE/ACM Transactions on Networking*, vol. 12, no. 3, pp. 493–506, 2004.

[YIC 08] YICK J., MUKHERJEE B., GHOSAL D., "Wireless sensor network survey", *Computer Networks*, vol. 52, no. 12, pp. 2292–2330, 2008.

[YOU 04] YOUNIS O., FAHMY S., "Heed: a hybrid, energy-efficient, distributed clustering approach for ad-hoc networks", *IEEE Transactions on Mobile Computing*, vol. 3, no. 4, pp. 366–369, 2004.

[YU 01] YU Y., ESTRIN D., GOVINDAN R., Geographical and energy-aware routing: a recursive data dissemination protocol for wireless sensor networks, Technical Report UCLA-CSD TR-01-0023, Computer Science Department, UCLA, May 2001.

[ZHA 04] ZHAO W., AMMAR M., ZEGURA E., "A message ferrying approach for data delivery in sparse mobile ad hoc networks", *Proceedings of the 5th ACM International Symposium on Mobile Ad Hoc Networking and Computing*, Tokyo, Japan, pp. 187–198, May 2004.

[ZHE 03] ZHENG R., HOU J.C., SHA L., "Asynchronous wakeup for ad hoc networks", *Proceedings of the 4th ACM International Symposium on Mobile Ad Hoc Networking and Computing*, pp. 35–45, 2003.

[ZHU 03] ZHU J., PAPAVASSILIOU S., "On the energy-efficient organization and the lifetime of multi-hop sensor networks", *IEEE Communication Letter*, vol. 7, no. 11, pp. 537–539, 2003.

[ZOR 03a] ZORZI M., RAO R.R., "Geographic random forwarding (GeRaF) for adhoc and sensor networks: multihop performance", *IEEE Transactions on Mobile Computing*, vol. 2, no. 4, pp. 337–348, 2003.

[ZOR 03b] ZORZI M., RAO R.R., "Geographic random forwarding (GeRaF) for ad hoc and sensor networks: energy and latency performance", *IEEE Transactions on Mobile Computing*, vol. 2, no. 4, pp. 349–365, 2003.

Index

A, C, D, E

M, R, S, T

Printed in the United States
By Bookmasters